# Drilling Fluids Processing Handbook

# Drilling Fluids Processing Handbook

Editor

**Dipti Mishra**

**Drilling Fluids Processing Handbook**

Edited by **Dipti Mishra**

Printed in 2017

ISBN: 978-1-68117-356-6

Library of Congress Control Number: 2015941548

© 2016 by
SCITUS Academics LLC,
616, Corporate Way, Suite 2, 4766,
Valley Cottage, NY 10989

www.scitusacademics.com

This book contains information obtained from highly regarded resources. Copyright for individual articles remains with the authors as indicated. All chapters are distributed under the terms of the Creative Commons Attribution License, which permits unrestricted use, distribution, and reproduction in any medium, provided the original author and source are credited.

**Notice**

Reasonable efforts have been made to publish reliable data and views articulated in the chapters are those of the individual contributors, and not necessarily those of the editors or publishers. Editors or publishers are not responsible for the accuracy of the information in the published chapters or consequences of their use. The publisher believes no responsibility for any damage or grievance to the persons or property arising out of the use of any materials, instructions, methods or thoughts in the book. The editors and the publisher have attempted to trace the copyright holders of all material reproduced in this publication and apologize to copyright holders if permission has not been obtained. If any copyright holder has not been acknowledged, please write to us so we may rectify.

# Contents

Preface ............................................................................................................. vii

Chapter 1　Calculation Analysis of Pressure Wave Velocity in Gas and Drilling Mud Two-Phase Fluid in Annulus during Drilling Operations ............................................................................................. 1

　　　　　　Yuanhua Lin, Xiangwei Kong, Yijie Qiu, and Qiji Yuan

Chapter 2　Diffusion of Chemically Reactive Species in Casson Fluid Flow over an Unsteady Stretching Surface in Porous Medium in the Presence of a Magnetic Field ................................................................................... 49

　　　　　　Gilbert Makanda, Sachin Shaw, and Precious Sibanda

Chapter 3　Encapsulation of Menthol in Beeswax by a Supercritical Fluid Technique ......................................................... 77

　　　　　　Linjing Zhu, Hongqiao Lan, Bingjing He, Wei Hong, and Jun Li

Chapter 4　Direct Solid-State Fermentation of Soybean Processing Residues for the Production of Fungal Chitosan by Mucor Rouxii ................................................................. 97

　　　　　　Andro Mondala, Ramea Al-Mubarak, James Atkinson, Shaun Shields, Brian Young, Yurguen Dos Santos Senger, and Jan Pekarovic

Chapter 5　A Promising Material by Using Residue Waste from Bisphenol a Manufacturing to Prepare Fluid-Loss-Control Additive in Oil Well Drilling Fluid ............................................... 119

　　　　　　Zhi-Lei Zhang, Feng-Shan Zhou, Yi-He Zhang, Hong-Wei Huang, Ji-Wu Shang, Li Yu, Hong-Zhen Wang, and Wang-Shu Tong

Chapter 6　Evaluation of Additives as Corrosion Inhibitors/ Antioxidants for High Quality Nano Emulsifiable Oils of Metalworking Fluids ....................................................................... 147

　　　　　　Noura El Mehbad

| | | |
|---|---|---|
| Chapter 7 | Effect of Processing Paramters on Metal Matrix Composites: Stir Casting Process .......................................................................... 163 | |

G. G. Sozhamannan, S. Balasivanandha Prabu, and V. S. K. Venkatagalapathy

| | | |
|---|---|---|
| Chapter 8 | The Effect of Drilling Fluids and Crude Oil on Some Chemical Characteristics of Soil and Crops .................................................... 177 | |

Ivica Kisic, Sanja Mesic, Ferdo Basic, Vladislav Brkic, Milan Mesic, Goran Durn, Zeljka Zgorelec, Lidija Bertovic

| | | |
|---|---|---|
| Chapter 9 | Influence of Viscosity Modifier Nature and Concentration on the Viscous Flow Behaviour of Oil-Based Drilling Fluids at High Pressure ....................................................... 207 | |

Ivica Kisic, Sanja Mesic, Ferdo Basic, Vladislav Brkic, Milan Mesic, Goran Durn, Zeljka Zgorelec, Lidija Bertovic

Citations ......................................................................................... 235

Index ............................................................................................... 239

# Preface

Drilling fluid is used to aid the drilling of boreholes into the earth. Often used while drilling oil and natural gas wells and on exploration drilling rigs, drilling fluids are also used for much simpler boreholes, such as water wells. Liquid drilling fluid is often called drilling mud. The three main categories of drilling fluids are water-based muds (which can be dispersed and non-dispersed), non-aqueous muds, usually called oil-based mud, and gaseous drilling fluid, in which a wide range of gases can be used. The main functions of drilling fluids include providing hydrostatic pressure to prevent formation fluids from entering into the well bore, keeping the drill bit cool and clean during drilling, carrying out drill cuttings, and suspending the drill cuttings while drilling is paused and when the drilling assembly is brought in and out of the hole. The drilling fluid used for a particular job is selected to avoid formation damage and to limit corrosion.

**Editor**

# Calculation Analysis of Pressure Wave Velocity in Gas and Drilling Mud Two-Phase Fluid in Annulus during Drilling Operations

Yuanhua Lin[1], Xiangwei Kong[2], Yijie Qiu[2], and Qiji Yuan[1]

[1]State Key Laboratory of Oil and Gas Reservoir Geology and Exploitation, Southwest Petroleum University, Chengdu, Sichuan 610500, China

[2]School of Petroleum Engineering, Southwest Petroleum University, Chengdu, Sichuan 610500, China

## ABSTRACT

Investigation of propagation characteristics of a pressure wave is of great significance to the solution of the transient pressure problem caused by unsteady operations during management pressure drilling operations. With consideration of the important factors such as virtual mass force, drag force, angular frequency, gas influx rate, pressure, temperature, and well depth, a united wave velocity model has been proposed based on pressure gradient equations in drilling operations, gas-liquid two-fluid model, the gas-drilling mud equations of state, and small perturbation theory. Solved by adopting the Runge-Kutta method, calculation results indicate that the wave velocity and void fraction have different values with respect to well depth. In the annulus, the drop of pressure causes an increase in void fraction along the flow direction. The void fraction increases first slightly and then sharply; correspondingly the wave velocity first gradually decreases and then slightly increases. In general, the wave velocity tends to increase with the increase in back pressure and the decrease of gas influx rate and angular frequency, significantly in low range. Taking the virtual mass force into account, the dispersion characteristic of the pressure wave weakens obviously, especially at the position close to the wellhead.

# INTRODUCTION

One of the future trends of the petroleum industry is the exploration and development of high pressure, low permeability reservoirs [1]. Drilling-related issues such as excessive mud cost, wellbore ballooning/breathing, kick-detection limitations, difficulty in avoiding gross overbalance conditions, differentially stuck pipe, and resulting well-control issues together contribute to the application of managed pressure drilling (MPD) technology [2]. MPD technology has the ability to quickly react to the expected drilling problems and formations pressures uncertainties, reduce nonproductive time and mitigating drilling hazards, and offer a considerable amount of tangible benefits while drilling in extremely narrow fracture/pore pressure windows [3]. It allows drilling operations to proceed where conventional drilling is easy to cause formation damage or considered uneconomical, of high risk, or even impossible [4]. Although drilling operators try to avoid the risk of influxes, occasionally there are influxes for various reasons. Gas influx occurs whenever the pressure of a gas-bearing formation exceeds the pressure at the bottom of a wellbore. Since the subsequent intrusion of gas displaces drilling mud, it decreases the pressure in the wellbore and makes gas enter even faster [5, 6]. If it is not counteracted in time, the unstable effect can escalate into a blowout creating severe financial losses, environmental contamination, and potential loss of human lives. The basic principle of MPD well control is to keep the bottomhole pressure (BHP) as constant as possible at a value that is at least equal to the formation pressure [7]. MPD is a class of techniques that allow precise management of BHP under both static and dynamic conditions through a combination of controlling the flow rate, mud density, and back pressure (or wellhead pressure) on the fluid returns (choke manifold) of the enclosed and pressurized fluid system [8, 9]. As a result of the ability to manipulate the back pressure, MPD offers the capability to improve safety and well control through early detection of influxes and losses in a microflux level, reduces the risk of influx and thus the chance of blowouts, and controls an influx dynamically without conventionally shutting-in [10].

During the managed pressure drilling process, all the unsteady operations such as adjustment of choke, wellhead back pressure controls [11], tripping in/out [12], shutting-in [13], and mud pump rate changes [14] will cause generation and propagation of pressure waves, which would threaten the whole drilling system from the wellhead facilities to the bottomhole drilling equipment and the formation [15]. In the analysis of an influx well and formulation of well control scheme, the dynamic effects of these operations, appeared as pressure fluctuations, should be accounted for [16]. As a basic parameter of pressure fluctuation, the pressure wave velocity has close relevance with the determination of transient pressure and safe operating parameters for well control. However, the gas influx is more troublesome for the higher compressibility and lower density of influx gas than the single phase drilling mud [17]. The works of Bacon et al. [18] had demonstrated that compressibility effects of a gas influx can be significant during an applied-back-pressure, dynamic, MPD well control response and can impact the well control process. Hence, this paper considered the propagation behavior of pressure wave in gas-drilling mud two-phase flow in the annulus to provide reference for the MPD operations.

Well control includes not only the handling but also early detection of a gas influx. Besides the transient pressure problem mentioned previously, investigation of the propagation characteristics of pressure wave is of great significance to the early detection of gas influx [19]. Many scholars believe that the pressure fluctuations procedure contains a wealth of information about the flow. Thus, characteristics of pressure wave can be easily used to measure some important parameters in the two-phase flow [20]. In the late 1970s, the former Soviet All-Union Drilling Technology Research Institute began to study characteristics of pressure wave propagation velocity in gas-liquid two-phase flow to detect early gas influx and achieved some important results [21]. According to the functional relationship between the pressure wave propagation velocity in gas-liquid two-phase flow and the gas void fraction, Li et al. [22] presented a method of detecting the gas influxes rate and the height of gas migration early after gas influxes into the wellbore. Furthermore, mud pulse telemetry is the most common method of data transmission used by measurement-while-drilling, and the transmission velocity of the pulse is a basic parameter for this kind of data transmission mode [23].

The pressure wave discussed in this paper can be transmitted as a pressure perturbation along the direction of flow in wellbore, which propagates with the speed of sound in the mud and gas two-phase drilling fluid. Due to the compressibility of the gas phase, the changes in interface between the gas and drilling mud, and the momentum and energy transfer between two phases, it is complicated to predict the pressure wave velocity in gas and drilling mud two-phase flow. Since the 1940s, many experimental and theoretical studies have been performed. Experimental tests were conducted to inspect the contributions of fluctuation and flow characteristics on pressure wave. Ruggles et al. [24] firstly performed the experimental investigation on the dispersion property of pressure wave propagation in air-water bubbly flow. It was demonstrated that the propagation speed of pressure wave varies over a range of values for the given state, depending on the angular frequency of the pressure wave. Legius et al. [25] tested the propagation of pressure waves in bubbly and slug flow. The experimental result is similar to the calculation result of the Nguyen model and simulation result of the Sophy-2 package. Concluded form Miyazaki and Nakajima experiments [26] in Nitrogen-Mercury two-phase system, the slip between the phases plays a very important role in the mechanism of pressure wave propagation. From experimental investigation, Bai [27] found that the fractal dimension, correlation dimension, and the Kolmogorov entropy have close relationship with flow regime, and the fractal dimension will be greater than 1.5 when the flow is annular with high gas velocity. The characteristics of pressure wave propagation in bubbly and slug flow in a vertical pipe were investigated experimentally by Huang et al. in detail [28]. It confirmed that the propagation velocity is greatly affected by the gas void fraction and angular frequency of the pressure disturbance, and the superficial velocity of flowing medium has almost no effect on the propagation velocity. Also, there are some widely accepted models including elasticity model, homogeneous mixture model, and continuum model for pressure wave propagation in gas-liquid two-phase flow. Tangren et al. [29] took the two-phase media as a homogeneous fluid and presented the solution model concerning the problem of pressure wave velocity in two-phase flow at low gas void fraction. Wallis [30] studied the propagation mechanism of pressure waves and derived the propagation velocity in a bubbly flow and separated flow using the homogeneous model, in which the two-phase mixture is treated as a compressible fluid with suitably averaged properties. Nguyen et al. [31] applied the elastic theory to predict the propagation velocity of pressure waves in several different flow regimes. The comparison between the calculation results and available experimental data suggests its success at low void fraction. Mecredy and Hamilton [32] derived a detailed continuum model for sound wave propagation in gas-liquid flow by using six separate conservation equations to describe the flow of the vapor and liquid phases. This so-called two-fluid representation allows for nonequilibrium mass, heat, and momentum transfer between the phases. Results indicated that in a bubbly flow, high angular frequency waves travelled an order of magnitude faster than low angular frequency waves. With the development of hydrodynamics, the two-fluid model is widely used to determine the propagation velocity of the pressure wave in two-phase flow as it can provide a general dispersion relation valid for arbitrary flow regimes including effects of the interphase mass, heat, and momentum transfer. Ardron and Duffey [33] developed a model for sound-wave propagation in nonequilibrium vapor-liquid flows which predicts sound speeds and wave attenuations dependent only on measurable flow properties on the basis of two-fluid conservation equations. Ruggles et al. [24], Xu and Chen [34], Chung et al. [35], Huang et al. [28], and Bai et al. [27] investigated the propagation velocity behavior of pressure wave via two-fluid model and small perturbation theory, and the predicted results show good agreement with the experimental data. In recent years, some new researches and models that are especially important in this area were developed. Xu and Gong [36] proposed a united model to predict wave velocity for different flow patterns. In this united model, the effect of a virtual mass coefficient was taken into consideration. The propagation of pressure wave during the condensation of R404A and R134a refrigerants in pipe mini-channels that undergo periodic hydrodynamic disturbances was given by Kuczynski [37, 38]. Li et al. [39] simulated the condensation of gas oxygen in subcooled liquid oxygen and the corresponding mixing process in pump pipeline with the application of thermal phase change model in Computational Fluid Dynamics code CFX and investigated the pressure wave propagation characteristics in two-phase flow pipeline for liquid-propellant rocket based on a proposed pressure wave propagation model and the predicted flow parameters. Based on the unified theory of Kanagawa et al. [40], the nonlinear wave equation for pressure wave propagation in liquids containing gas bubbles is rederived. On the basis of numerical simulation of the gaseous oxygen and liquid oxygen condensation process with the thermal phase model in ANSYS CFX, Chen et al. [41] solved the pressure wave propagation velocity in pump pipeline via the dispersion equation derived from ensemble average two-fluid model. Li and He [42] developed an improved slug flow tracking model and analyzed the variation rule of the pressure wave along the pipeline and influence of the variation of initial inlet liquid flow rate and gas flow rate in horizontal air-water slug flow with transient air flow rate. Meanwhile, the compressibility effects of gas had been noticed in the research field of propagation of pressure waves in the drilling industry, and some efforts have been made. Li et al. [22] established the relationship between wave velocity and gas void fraction according to the empirical formula of the homogeneous mixture model presented by Martin and Padmanatbhan [43] and frequency response model presented by Henry [44]. By applying the unsteady flow dynamic theory, Liu et al. [45] derived the pressure wave velocity calculation formula for gas-drilling mud-solid three-phase flow based on the continuity equation. Wang and Zhang [46] studied the pressure pulsation in mud and set up a model for calculating the amplitude of pressure pulsation when pressure wave is transmitted in drilling-fluid channel especially drilling hose with different inside diameters. However some efforts have been made; the pressure wave velocity is usually determined by empirical formula. In the past researches, the influencing factors for pressure wave propagation were simulated and analyzed with the mathematical model; however, the variation of wave velocity and gas void at different depth of wellbore was not considered. In addition, the current researches are limited in their assumption and neglect the flow pattern translation and interphase forces along the annulus. Up to now, no complete mathematical model of pressure wave in an annulus with variations of gas void, flow pattern, temperature, and back pressure during MPD operations has been derived.

The object of the present work is to study the velocity for the transmission of pressure disturbance in the two-phase drilling fluid in the form of a pressure wave in annulus during MPD operations. In this paper, in addition to the pressure, temperature, and the void fraction in the annulus, the compressibility of the gas phase, the virtual mass force, and the changes of interface in two phases are also taken into consideration. By introducing the pressure gradient equations in MPD operations, gas-liquid two-fluid model, the gas-drilling mud equations of state (EOS), and small perturbation theory, a united model for predicting pressure wave velocity in gas and drilling mud in an annulus is developed. The model can be used to predict the wave velocity of various annulus positions at different influx rates, applied back pressures, and angular frequencies with a full consideration of drilling mud compressibility and interphase forces.

# THE MATHEMATICAL MODEL

## The Basic Equation

In this paper, the two-fluid model and the pressure gradient equation along the flow direction in the annulus are combined to study the pressure wave velocity in MPD operations. Drilling fluid contains clay, cuttings, barite, other solids, and so forth. The solid particles are small and uniformly distributed; therefore, drilling fluid is considered to be a pseudohomogeneous liquid, and the influx natural gas is considered to be the gas phase. The following assumptions are made:
- the two-phase flow is treated as one dimension;
- no mass transfers between the gas and drilling mud;
- the flow pattern in an annulus is either bubbly or slug flow.

As shown in Figure 1, the gas and drilling mud two-phase fluid travels along the annulus in the drilling process. The fluid flows along the annulus in "-z" direction, and the annulus is formed by the casing and drill string.

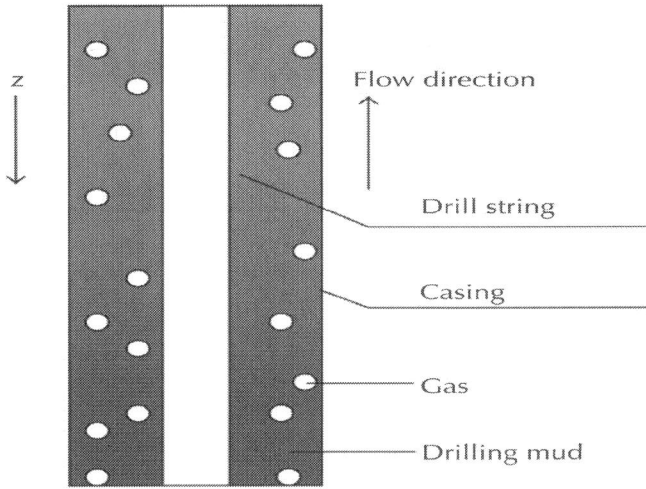

**Figure 1:** The schematic of gas and drilling mud two-phase flow in annulus.

The momentum conservation equation for gas and liquid two-phase flow is

$$\sum F_z = \rho_m A dz \frac{dv_m}{dt}. \tag{1}$$

The mass conservation equations are

$$\frac{\partial}{\partial t}(\phi_G \rho_G) + \frac{\partial}{\partial z}(\phi_G \rho_G v_G) = 0,$$
$$\frac{\partial}{\partial t}(\phi_L \rho_L) + \frac{\partial}{\partial z}(\phi_L \rho_L v_L) = 0. \tag{2}$$

The momentum conservation equations for gas are

$$\frac{\partial}{\partial t}(\phi_G \rho_G v_G) + \frac{\partial}{\partial z}(\phi_G \rho_G v_G^2) + \frac{\partial}{\partial z}(\phi_G \rho_G)$$

$$-\frac{\partial}{\partial z}\left[\phi_G\left(\tau_G^{fr} + \tau_G^{Re}\right) + M_{Gi} - 4\frac{\tau_G}{D}\right] = 0.$$

(3)

The momentum conservation equations for liquid are

$$\frac{\partial}{\partial t}(\phi_L \rho_L v_L) + \frac{\partial}{\partial z}(\phi_L \rho_L v_L^2) + \frac{\partial}{\partial z}(\phi_L \rho_L)$$

$$-\frac{\partial}{\partial z}\left[\phi_L\left(\tau_L^{fr} + \tau_L^{Re}\right) + M_{Li} - 4\frac{\tau_L}{D}\right] = 0.$$

(4)

The interphase forces include virtual mass force, drag force, and the wall shear stress. The transfer of momentum between the gas and drilling mud phases $M_{Gi}$ and $M_{Li}$ can be written as follows:

$$M_{Gi} = -M_{Li}^{nd} - M_{Li}^{d} + \left(\tau_{Li}^{fr} + \tau_{Li}^{Re}\right)\frac{\partial \phi_L}{\partial z}$$

$$+ \frac{\partial(\phi_G \sigma_s)}{\partial z} + \frac{\partial(\phi_G p_{Gi})}{\partial z} - \phi_G \frac{\partial(p_{Li})}{\partial z},$$

$$M_{Li} = M_{Li}^{nd} + M_{Li}^{d} + p_{Li}\frac{\partial(\phi_L)}{\partial z} - \left(\tau_{Li}^{fr} + \tau_{Li}^{Re}\right)\frac{\partial \phi_L}{\partial z}.$$

(5)

The momentum transfer term caused by the gas-liquid interfacial relative acceleration motion (i.e., the virtual mass force) can be expressed as

$$M_{Li}^{nd} = c_{VM}\phi_G \rho_L \alpha_{VM} - 0.1\phi_G \rho_L v_s \frac{\partial v_s}{\partial z} - c_{m1}\rho_L v_s^2 \frac{\partial \phi_G}{\partial z},$$

(6)

where $c_{m1} = 0.1$ and $vs$ is the slip velocity; $Vs = V_G - V_L$

The momentum transfer term caused by the drag force provided by Park et al. [47] can be described as

$$M_{Li}^d = \frac{3}{8}\frac{C_D}{r}\rho_L R_G v_s^2. \tag{7}$$

The pressure difference between the liquid interface and liquid is defined by the following formula presented by Park et al.:

$$p_{Li} - p_L = -c_p \rho_L v_s^2, \tag{8}$$

where $C_p = 0.25$.

The proportion of gas phase in the interface between the gas and drilling mud is rather small in that the pressure difference between the gas interface and gas is not very high. Omitting the pressure difference, the gas interface pressure $P_{Gi}$ can be written as

$$p_{Gi} - p_G \approx 0. \tag{9}$$

According to the equations of Arnold [48], the pressure of the drilling mud is described as follows:

$$p_L = p - 0.25\rho_L \phi_G v_s^2. \tag{10}$$

In fact, the shear stress and the interphase shear stress are very small relative to the Reynolds stress. Also, the Reynolds stress in gas phase can be omitted relative to the interphase force. Hence, it can be described as

$$\tau_G^{fr} \approx \tau_{Li}^{fr} \approx \tau_L^{fr} \approx \tau_G \approx \tau_G^{Re} \approx 0. \tag{11}$$

The Reynolds stress and interfacial average Reynolds stress can be obtained by

$$\tau_L^{Re} = -c_r \rho_L v_s^2 \frac{\phi_G}{\phi_L},$$

$$\tau_{Li}^{Re} = -c_r \rho_L v_s^2, \tag{12}$$

where $C_r = 0.2$.

The wall shear stress of liquid phase is given by Wallis in the following form [30]:

$$\tau_L = 0.5 f_L \rho_L v_L^2. \tag{13}$$

## Physical Equations

### *Equations of State for Gas*

The equation of state (EOS) for gas can be expressed as follows:

$$\rho_G = \frac{p}{(Z_G \cdot R \cdot T)}, \tag{14}$$

where $Z_G$ is the compression factor of gas.

The formula presented by Dranchuk and Abou-Kassem has been used to solve the gas deviation factor under the condition of low and medium pressure ($p < 35$ MPa) [49]:

$$Z_G = 1 + \left(0.3051 - \frac{1.0467}{T_r} - \frac{0.5783}{T_r^3}\right)\rho_r$$

$$+ \left(0.5353 - \frac{02.6123}{T_r} - \frac{0.6816}{T_r^3}\right)\rho_r^2, \tag{15}$$

where $T_r = T/T_c$, $pr = p/p_c$, and $\rho_r = 0.27 p_r/Z_G T_r$.

The formula presented by Yarborough and Hall has been adopted to solve the gas deviation factor under the condition of high pressure ($p \geq 35$ MPa) [50]:

$$Z_G = \frac{0.06125 P_r T_r^{-1} \exp\left(-1.2\left(1 - T_r^{-1}\right)2\right)}{Y},\tag{16}$$

where Y is given by

$$\begin{aligned}
-0.06125 p_r T_r^{-1} &\exp\left(-1.2\left(1 - T_r^{-1}\right)^2\right) + \frac{Y + Y^2 + Y^3 + Y^4}{(1-Y)^3} \\
&= \left(14.76 T_r^{-1} - 9.76 T_r^{-2} + 4.58 T_r^{-3}\right) Y^2 \\
&\quad - \left(90.7 T_r^{-1} - 242.2 T_r^{-2} + 42.4 T_r^{-3}\right) Y^{(2.18 + 2.82 T_r^{-1})}.
\end{aligned}\tag{17}$$

## Equations of State for Liquid

Under different temperatures and pressures, the density of drilling mud can be obtained by the empirical formulas.

If $T < 130\,°C$, the density of drilling mud can be obtained by the following equation:

$$\rho_L = \rho_0 \left(1 + 4 \times 10^{-10} p_L - 4 \times 10^{-5} T - 3 \times 10^{-6} T^2\right).\tag{18}$$

If $T > 130\,°C$, the density of drilling mud is

$$\rho_L = \rho_0 \left(1 + 4 \times 10^{-10} p_L - 4 \times 10^{-5} T \right.\\
\left. -3 \times 10^{-6} T^2 + 0.4 \left(\frac{T - 130}{T}\right)^2\right).\tag{19}$$

## Correlation of Temperature Distribution

The temperature of the drilling mud at different depths of the annulus can be determined by the relationship presented by Hasan and Kabir [51]:

$$T = T_{ei} + F \left[1 - e^{(z_{bh}-z)/A}\right] \left(-\frac{g \sin \theta}{g_c J c_{pm}} + N + g_T \sin \theta\right)$$

$$+ e^{(z_{bh}-z)/A} \left(T_{fbh} - T_{ebh}\right). \qquad (20)$$

# Flow Pattern Analysis

Based on the analysis of flow characteristics in the enclosed drilling system, it can be safely assumed that the flow pattern in an annulus is either bubbly or slug flow [52]. The pattern transition criteria for bubbly flow and slug flow given by Orkiszewski are used to judge the flow pattern in the gas-drilling mud two-phase flow [53].

For bubbly flow,

$$\frac{q_G}{q_m} < L_B. \qquad (21)$$

For slug flow,

$$\frac{q_G}{q_m} > L_B, \qquad N_{GV} < L_S, \qquad (22)$$

where $q_m$ is the volumetric flow rate of two-phase flow, $q_m = q_G + q_L$.

The dimensionless numbers $L_S$ and $L_B$ are defined as

$$L_S = 50 + 36 N_{GV} \frac{q_L}{q_G},$$

$$L_B = 1.071 - \frac{0.7277 v_m^2}{D}, \tag{23}$$

Where

$$N_{GV} = v_s \left(\frac{\rho_L}{g \sigma_s}\right)^{0.25}. \tag{24}$$

Flow parameters such as void fraction, mixture density, and virtual mass force coefficient are discussed for a specific flow pattern.

The correlation between void fraction and liquid holdup is expressed as

$$\phi_G + \phi_L = 1. \tag{25}$$

## Bubbly Flow

As for bubbly flow, the density of gas and drilling mud mixture is described as the following formula:

$$\rho_m = \phi_L \rho_L + \phi_G \rho_G. \tag{26}$$

The void fraction is determined by the following formula:

$$\phi_G = \frac{1}{2}\left[1 + \frac{q_m}{v_s A} - \sqrt{\left(1 + \frac{q_m}{v_s A}\right)^2 - \frac{4q_G}{v_s A}}\right]. \tag{27}$$

The coefficient of virtual mass force $C_{vm}$ for bubbly flow can be expressed as follows [54]:

$$C_{vm} = 0.5\frac{1 + 2\phi_G}{1 - \phi_G}. \tag{28}$$

The coefficient of resistance coefficient $C_D$ for bubbly flow can be expressed as

$$C_D = \frac{4R_b}{3}\sqrt{\frac{g(\rho_L - \rho_G)}{\sigma_s}}\left[\frac{1 + 17.67(1-\phi_G)^{9/7}}{18.67(1-\phi_G)^{1.5}}\right]^2. \tag{29}$$

The friction pressure gradient for bubbly flow can be obtained from the following equation:

$$\tau_f = f\frac{\rho_L v_L^2}{2D}. \tag{30}$$

## Slug Flow

As for slug flow, distribution coefficient of gas in the liquid phase is

$$C_0 = \frac{0.00252\lg(10^3 \mu_L)}{A^{1.38}} - 0.782 + 0.232\lg v_m - 0.428\lg A. \tag{31}$$

The average density of the mixture for slug flow is determined by

$$\rho_m = \frac{W_m + \rho_L v_s A}{q_m + v_s A} + C_0 \rho_L. \tag{32}$$

The void fraction for slug flow is determined by the following formula:

$$\phi_G = \frac{q_G}{q_G + q_L}. \tag{33}$$

The coefficient of virtual mass force $C_{vm}$ for slug flow can be expressed as follows:

$$C_{vm} = 3.3 + 1.7 \frac{3L_q - 3R_q}{3L_q - R_q}. \tag{34}$$

The coefficient of resistance coefficient $C_D$ for slug flow can be expressed as

$$C_D = 110 \phi_L^3 R_b. \tag{35}$$

The friction pressure gradient of bubbly flow can be obtained from the following equation:

$$\tau_f = f \frac{\rho_L v_m^2}{2D} \left( \frac{q_L + v_s A}{q_m + v_s A} + C_0 \right). \tag{36}$$

## Annulus Characteristic Analysis

The annulus effective diameter proposed by Sanchez [55] is used in the hydraulic calculation of annulus

$$D = \frac{\pi (D_o^2 - D_i^2)/4}{\pi (D_o + D_i)/4} = D_o - D_i. \tag{37}$$

The effective roughness of the annulus can be expressed as

$$k_e = k_o \frac{D_o}{D_o + D_i} + k_i \frac{D_i}{D_o + D_i}, \tag{38}$$

where the $D_i$ and $D_o$ are the diameters of inner pipe and outer pipe, respectively; the $K_o$ and $K_i$ are the roughnesses of outer pipe and the inner pipe, respectively.

# THE UNITED MODEL DEVELOPED

The total pressure drop gradient is the sum of pressure drop gradients due to potential energy change and kinetic energy and frictional loss. From (1), the equation used to calculate the pressure gradient of gas and drilling mud flow within the annulus can be written as

$$\frac{dp}{dz} = \rho_m g \sin \theta - \frac{\tau_w \pi D}{A} - \rho_m v_m \frac{dv_m}{dz}. \tag{39}$$

Assuming the compressibility of the gas is only related to the pressure in the annulus, the kinetic energy or acceleration term in the previous equation can be simplified to

$$\rho_m v_m \frac{dv_m}{dz} = -\frac{\rho_m v_m v_{sg}}{p}\frac{dp}{dz} = -\frac{W_m q_g}{A^2 p}\frac{dp}{dz}. \tag{40}$$

Substituting (39) into (40), the total pressure drop gradient along the flow direction within the annulus can be expressed as

$$\frac{dp}{dz} = \frac{\rho_m g + \tau_f}{1 - W_m q_G/(A^2 p)}. \tag{41}$$

It is assumed that the gas obeys to the EOS (14), and the compressibility of drilling mud can be obtained by adopting the simplified EOS ((18) and (19)) which neglects the thermal expansion of liquid. The sonic speed of gas phase $C_G$ and that of liquid phase $C_l$ can be presented in the following form:

$$\frac{dp_L}{d\rho_L} = c_L^2,$$

$$\frac{dp_G}{d\rho_G} = c_G^2. \tag{42}$$

By introducing (42), the hydrodynamic equations of two-fluid model ((2)–(4)) can be written in the matrix form

$$A\frac{\partial \xi}{\partial t} + B\frac{\partial \xi}{\partial z} = C\xi, \tag{43}$$

where A is the matrix of parameters considered in relation to time, B is the matrix of parameters considered in relation to the spatial coordinate, and C is the vector of extractions.

By introducing the small disturbance theory, the disturbance of the state variable $(\phi_G, p, V_G, V_L)^T$ can be written as

$$\xi = \xi_0 + \delta\xi \exp\left[i\left(wt - kt\right)\right]. \tag{44}$$

In (44), K is the wave number, and w is the angular frequency of the disturbances. Substituting (44) into (43) gives homogenous linear equations concerning the expression n $(\delta\phi_G, \delta_p, \delta V_G, \delta V_L)^T$. According to the solvable condition of the homogenous linear equations that the determinant of the equations is zero, dispersion equation of pressure wave can be expressed in the following form:

$$\begin{vmatrix} M_1 & M_2 & M_3 & M_4 \\ -\rho_L w & \dfrac{1-\phi_G}{c_L^2} w & 0 & -k(1-\phi_G)\rho_L \\ M_5 & M_6 & M_7 & M_8 \\ M_9 & M_{10} & M_8 & M_{11} \end{vmatrix} = 0, \tag{45}$$

where $M_1$–$M_{11}$ can be illustrated by

$$M_1 = \left(\rho_G + c_p \phi_G \rho_L \dfrac{v_s^2}{c_G^2}\right) w,$$

$$M_2 = \dfrac{\phi_G}{c_G^2}\left[1 - c_p \phi_L\right] \dfrac{v_s^2}{c_L^2} w,$$

$$M_3 = -\left[\phi_G \rho_G k + 2c_p \phi_G \phi_L \rho_L \dfrac{v_s}{c_L^2} w\right],$$

$$M_4 = 22c_p\phi_G\phi_L\rho_L\frac{v_s}{c_L^2}w,$$

$$M_5 = \rho_L v_r^2 k\left(-\phi_G c_p + c_r - c_i + c_{m2}\right),$$

$$M_6 = -\phi_G k\left[1 - \phi_L\frac{c_p v_s^2}{c_L^2} + c_i\frac{v_s^2}{c_L^2}\right],$$

$$M_7 = \phi_G(\rho_G + c_{vm}\rho_L)w - i\left(\frac{3}{4}\frac{c_D}{r}\rho_L\phi_G v_s + \frac{4}{D}f_{Gw}\rho_G v_G\right),$$

$$M_8 = -c_{vm}\phi_G\rho_L w + i\left(\frac{3}{4}\frac{c_D}{r}\rho_L\phi_G v_s\right),$$

$$M_9 = \rho_L v_s^2 k\left(\phi_L c_p - 2c_r - c_{m2}\right),$$

$$M_{10} = -k\left(\phi_L + c_r\phi_G\frac{v_s^2}{c_L^2}\right),$$

$$M_{11} = \rho_L[\phi_L + \phi_G c_{vm}]w - i\left(\frac{3}{4}\frac{c_D}{r}\rho_L\phi_G v_s + \frac{4}{D}f_L\rho_L v_L\right), \quad (46)$$

where $c_{m2} = 0.1$, $c_p = 0.25$, $c_i = 0.3$, and $c_r = 0.2$.

By solving the dispersion equation indicated previously, four roots can be obtained. Because the wavelengths associated with two of the roots are too short to allow the two-fluid medium to be treated as a continuum, the two roots should be omitted. As for the two remaining roots, one of them represents a pressure wave that transmited along the axis z, and the other represents a pressure wave transmits in the opposite direction in accordance with the direction of the flow along the annulus. The real value of wave number can be used to determine the wave velocity c. The wave velocity in the two-phase fluid can be determined by the following model:

$$c = \frac{|(w/R^+(k)) - (w/R^-(k))|}{2}. \tag{47}$$

# SOLUTION OF THE UNITED MODEL

Obtaining the analytical solution of the mathematical models concerned with flow pattern, void fraction, characteristic parameters, and pressure drop gradient is generally impossible for two-phase flow. In this paper, the Runge-Kutta method (R-K4) is used to discretize the theoretical model.

We can obtain pressure, temperature, gas velocity, drilling mud velocity, and void fraction at different annulus depths by R-K4. The solution of pressure drop gradient equation (41) can be seen as an initial-value problem of the ordinary differential equation:

$$\frac{dp}{dz} = F(z, p),$$

$$p(z_0) = P_0. \tag{48}$$

With the initial value $(z_0, p_0)$ and the function F(z, p), (50)–(53) can be obtained by

$$k_1 = F(z_0, p_0), \tag{49}$$

$$k_2 = F\left(z_0 + \frac{h}{2}, p_0 + \frac{h}{2}k_1\right), \tag{50}$$

$$k_3 = F\left(z_0 + \frac{h}{2}, p_0 + \frac{h}{2}k_2\right), \tag{51}$$

$$k_4 = F(z_0 + h, p_0 + hk_3), \tag{52}$$

where h is the step of depth. The pressure on the nod $i = i + 1$ can be obtained by

$$p_1 = p_0 + \Delta p = p_0 + \frac{h}{6}(k_1 + 2k_2 + 2k_3 + k_4). \tag{53}$$

In the present work, the mathematical model and pressure wave velocity calculation model are solved by a personally compiled code on VB.NET (Version 2010). The solution procedure for the wave velocity in the annulus is shown in Figure 1. At initial time, the wellhead back pressure, wellhead temperature, wellbore structure, well depth, gas and drilling mud properties, and so forth, are known. On the node i, the pressure gradient, temperature, and the void fraction can be obtained by adopting R-K4. Then, the determinant (45) is calculated based on the calculated parameters. Omitting the two unreasonable roots, the pressure wave velocity at different depths of the annulus in MPD operations can be solved by (47). The process is repeated until the pressure wave velocity of every position in the wellbore is obtained as shown in Figure 2.

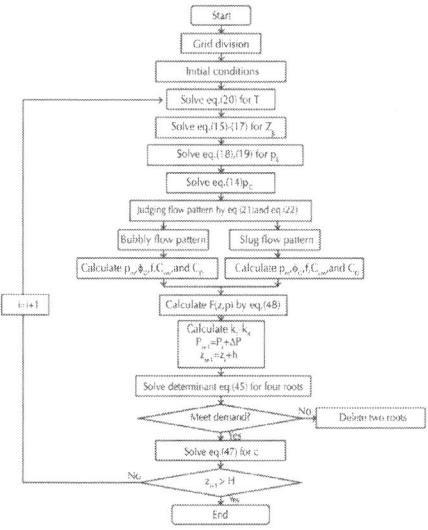

**Figure 2**: Solution procedure for wave velocity in MPD operations.

The developed model takes full consideration of the interfacial interaction and the virtual mass force. Owing to the complex conditions of the annulus in MPD operations, measurement of wave velocity in the actual drilling process is very difficult. In order to verify the united model, the predicted pressure waves are compared with the results of previous simulated experimental investigations presented for gas and drilling mud by Liu et al. [45] in Figure 3(a) and by Li et al. [22] in Figure 3(b). The lines represent the calculation results, and the points represent the experimental data.

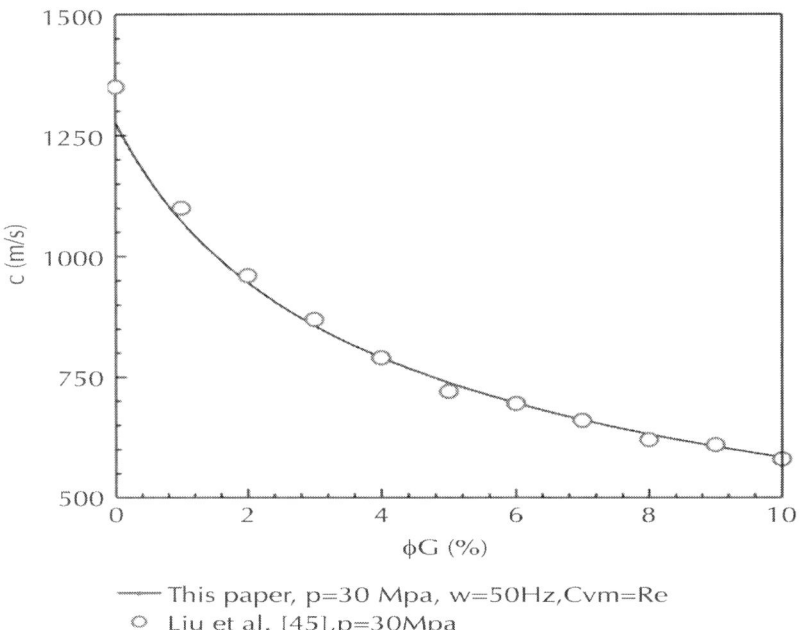

— This paper, p=30 Mpa, w=50Hz, Cvm=Re
○ Liu et al. [45], p=30Mpa

(a)Comparison with the experimental data of Liu

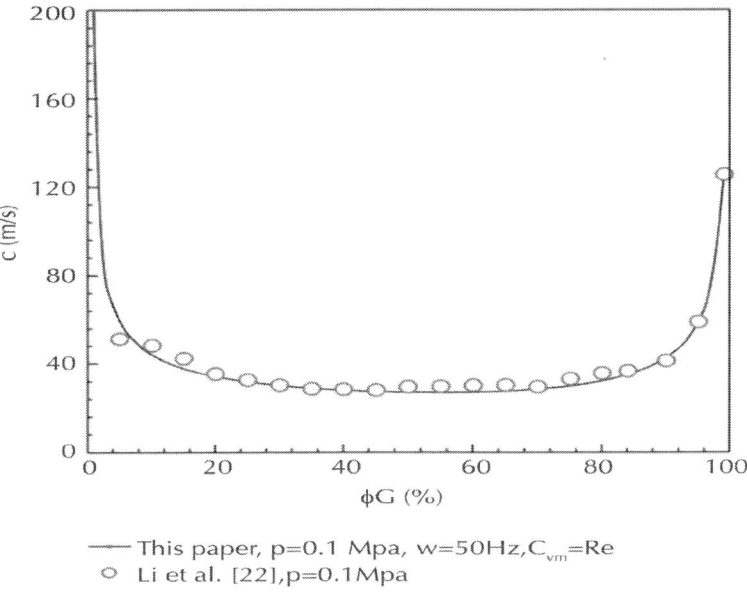

(b)Comparison with the experimental data of Li

**Figure 3:** Experimental verification by comparison with previous experimental data.

The comparisons reveal that the developed united model fits well with the experimental data. Thus, the united model can be used to accurately predict wave velocity at different wellhead back pressures and gas influx rates (the influx rate of gas in the bottomhole) in MPD operations.

## ANALYSIS AND DISCUSSION

The drilling system described is an enclosed system. The schematic diagram of the gas influx process is illustrated in Figure 4. The drilling mud is pumped from surface storage down the drill pipe. Returns from the wellbore annulus travel back through surface processing, where drilling solids are removed, to surface storage. The key equipments include the following.

- The Rotating Control Device (RCD). The rotating control device provides the seal between atmosphere and wellbore, while allowing pipe movement and diverting returns flow. In conjunction with the flow control choke, the RCD provides the ability to apply back pressure on the annulus during an MPD operation.
- Choke. The MPD choke manifold provides an adjustable choke system which is used to dynamically control the required BHP by means of applying surface BP.
- Coriolis Meter. A Coriolis meter is used to accurately measure the mass flow rate of fluid exiting the annulus. The ability to measure return flow accurately is essential for the applied back pressure.
- Pressure Sensor. A pressure sensor is used to measure surface back pressure on the wellhead.

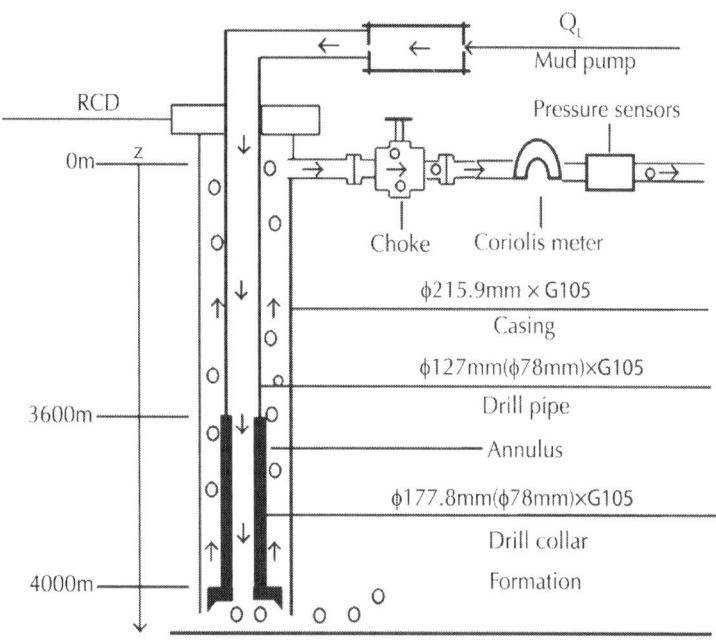

**Figure 4:** Schematic diagram of the gas influx process.

The gas and drilling mud flow rate measured by the Coriolis meter and the back pressure measured by the pressure sensor are the initial data for annulus pressure calculation. The well used for calculation is a gas well in Xinjiang Uygur Autonomous Region, Northwest China. The wellbore structure, well design parameters (depths and diameters), gas and drilling mud properties (density and viscosity), and operational conditions of calculation well are displayed in Table 1.

**Table 1:** Parameters of calculation well

| Type | Property | Value |
|---|---|---|
| Mud | Dynamic viscosity (Pa·s) | 0.056 |
| | Density (kg/cm$^3$) | 1460 |
| Gas | Relative density | 0.65 |
| | Viscosity (Pa·s) | $1.14 \times 10^{-5}$ |
| String | Elastic modulus of string (Pa) | $2.07 \times 10^{11}$ |
| | Poisson ratio of string | 0.3 |
| | Roughness (m) | $1.54 \times 10^{-7}$ |
| Surface condition | Surface temperature (K) | 298 |
| | Atmosphere pressure (MPa) | 0.101 |

The drilling mud mixed with gas is taken as a two-phase flow medium. The propagation velocity of pressure wave in the gas-drilling mud flow is calculated and discussed by using the established model and well parameters.

## Effect of Back Pressure on Wave Velocity

Increase in applied back pressure is the most common approach for dynamic well control. The wellbore can be seen as an enclosed pressurized system. The pressure at different depths of the annulus varied with the change of back pressure. According to the EOS of gas and drilling mud, the influence of pressure on gas volume is much greater than that of drilling mud for the greater compressibility of gas. So, the gas void fraction changes with the variation of gas volume at the nearly constant flow rate of drilling mud at different annular pressures. Meanwhile, the pressure wave velocity is sensitive to the gas void fraction. As a result, when the back pressure at the wellhead is changed by adjusting the choke, the wave velocity in the two-phase drilling fluid and the distribution of void fraction at different depth of the annulus will diverge. The calculation results affected by the back pressure are presented.

Figures 5 and 6 show the distributions of void fraction and variations of wave velocity along the flow direction in the annulus when the back pressure at the wellhead is 0.1 MPa, 0.8 MPa, 1.5 MPa, 2.5 MPa, 5.0 MPa, and 6.5 MPa, respectively. It can be seen that the void fraction significantly increases along the flow direction in the annulus. Conversely, the wave velocity shows a remarkable decrease tendency. For instance, BP=1.5 MPa, at the wellhead, the wave velocity is 90.12 m/s, and the void fraction is 0.611, while in the bottomhole, the wave velocity reaches 731.42 m/s, and void fraction is reduced to 0.059. However, a sudden increase in wave velocity is observed in Figure 5 when the back pressure is 0.1 MPa at the position close to wellhead as the void fraction is further increased. Moreover, the influence of back pressure on the void fraction is larger at the position close to the wellhead than that in the bottomhole.

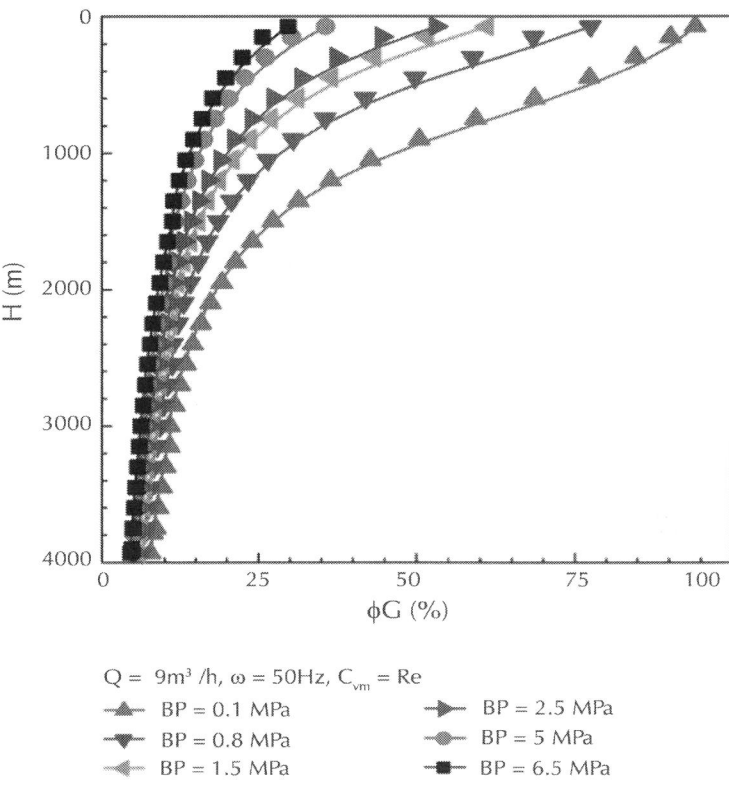

**Figure 5:** Void fraction distribution at different back pressures.

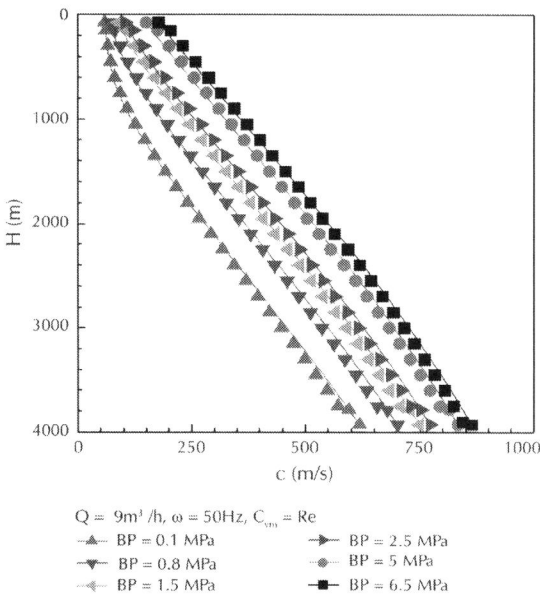

**Figure 6:** Wave velocity variations at different back pressures.

This can be explained from the viewpoints of mixture density and compressibility of two-phase fluid and the pressure drop along the flow direction in the wellbore. In the low void fraction range, the gas phase is dispersed in the liquid as bubble, so the wave velocity is influenced greatly by the added gas phase. According to the EOS, if gas invades into the wellbore with a small amount in the bottomhole, the density of the drilling mud has little variation while the compressibility increases obviously, which makes the wave velocity decreased. Then, the two-phase drilling mud flow from the bottomhole to the wellhead along the annulus with a drop of pressure caused by, potential energy change, kinetic energy and frictional loss, which leads to an increase in void fraction. When the void fraction is increased, both the density and the compressibility of the two-phase fluid change slightly, resulting in a flat decrease in wave velocity along the flow direction. With the further increasing in void fraction, the bubbly flow turns into slug flow according to the flow pattern transient criteria of Orkiszewski which shows that the flow pattern is dependent on the void fraction. Generally speaking, the gas-drilling mud slug flow is composed of liquid and gas slugs. In the preliminary slug flow, the liquid slugs are much longer than the gas slugs, so the wave velocity is determined primarily by the wave velocities in the liquid slugs which are hardly affected by the void fraction. At the position close to the wellhead, the pressure of two-phase flow fluid can be reduced to a low value which approaches the back pressure. The compressibility of gas will be improved at the low pressure. It results in significant increase in void fraction. As the void fraction increases greatly at the position close to the wellhead, gas slugs become much longer than the liquid slugs. It is assumed that the actual wave velocity in slug flow happens to play a leading role when the void fraction increases to some extent. For the decrease in liquid holdup in the gas slug with the increasing in void fraction, the wave velocity in the gas slug is infinitely close to that in pure gas, which is equivalent to an increase in wave velocity in the gas slug. Meanwhile, the gas content in liquid slug also increases and results in a lower wave velocity in the liquid slug. As a result, a slight increase in wave velocity appears.

Figures 7 and 8 present the influence of back pressure on the void fraction and wave velocity with respect to the parameters of well depth. As the back pressure increases, the void fraction at different depth of the annulus is reduced gradually, while the wave velocity in the two-phase flow tends to increase. Analytical results show that the increased back pressure is equivalent to be applied to the entire enclosed drilling fluid cyclical system. The pressure transmits from the wellhead to the bottomhole; therefore, the annular pressure in the entire wellbore is increased. According to the EOS, the density of gas increases and the compressibility of gas decreases with the increasing of gas pressure. So, the loss of interphase momentum and energy exchange is reduced and the interphase momentum exchange is promoted. It contributes to the increase in wave velocity with the increase in gas pressure. In addition, due to lower compressibility of two-phase flow medium under high pressure, the increase tendency of pressure wave velocity and the decrease tendency of void fraction are slowed down in the high back pressure range.

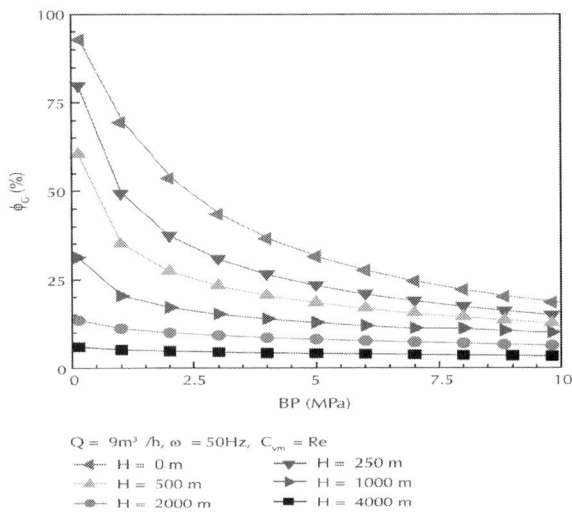

**Figure 7:** Effect of back pressure on the void fraction.

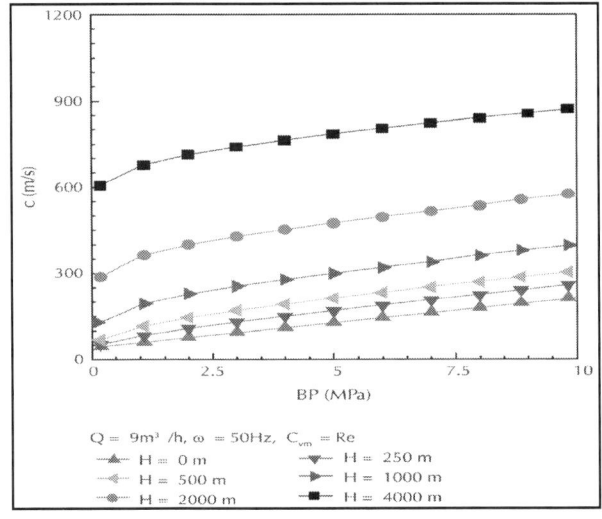

**Figure 8:** Effect of back pressure on the pressure wave velocity.

## Effect of Gas Influx Rate on Wave Velocity

Figures 9 and 10 graphically interpret the distributions of void fraction and variations of wave velocity along the flow direction in the annulus. When gas influx occurs in the bottomhole, gas invades into the wellbore and migrates from the bottomhole to the wellhead along the flow direction. At a low gas influx rate, it is extremely obvious that the void fraction and wave velocity first slightly change in a comparatively smooth value then change sharply. It is because of the rapid expansion of gas volume with the decreasing in pressure near the wellhead that the void fraction increases sharply, and the wave velocity decreases obviously at the same time. But under the high bottomhole pressure (up to 50 MPa), the compressibility of the gas is low. This results in a slight change in void fraction and wave velocity at the position far away from the wellhead. Since the compressible component increases with the increase in the gas influx rate, the compressibility of the gas and drilling mud two-phase fluid is improved. So the variations of void fraction and wave velocity become more prominent. Also, the void fraction still shows an increase tendency that is steady first and then sharp. It acts in accordance with the variation of void fraction at a low gas influx rate. At a low gas influx rate, if the void fraction at the position close to the wellhead can not increase to a high extent, the wave velocity always shows a decrease tendency. At a high gas influx rate, such as the $Q=8.312$ m$^3$/h, the wave velocity tends to increase because the void fraction in the wellhead is increased to a high extent. In conclusion, the wave pressure is sensitive to the void fraction, and the void fraction is dominated by influx rate and pressure in the annulus, especially the influx rate.

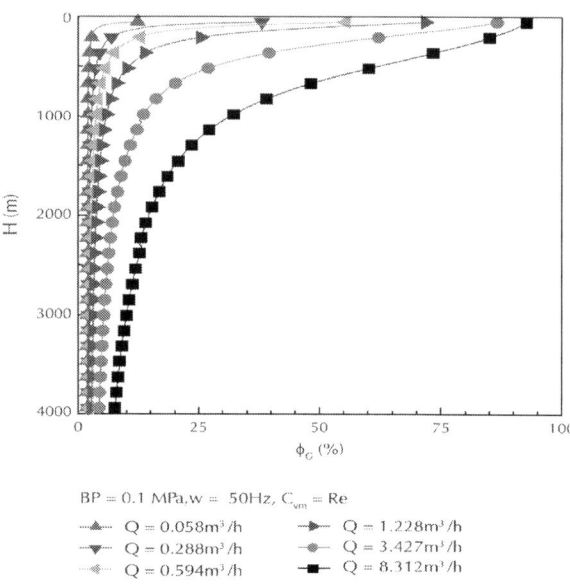

**Figure 9:** Void fraction distribution at different gas influx rates.

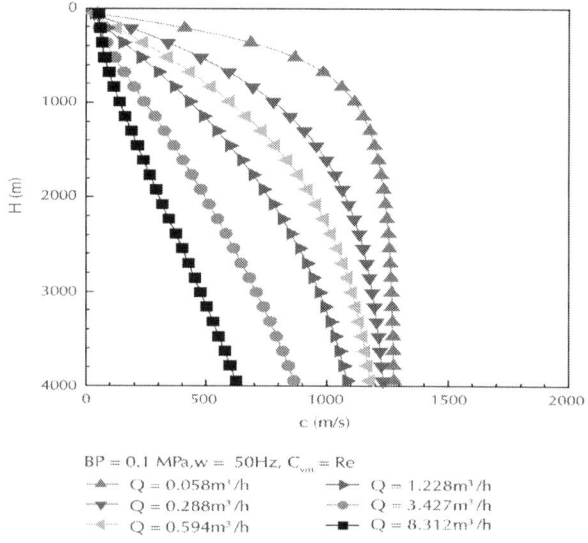

**Figure 10:** Wave velocity variations at different gas influx rates.

With the influx of gas, the mixing of gas and drilling mud occurs in the annulus and the corresponding interfacial transfer of momentum and mass causes an increase in gas phase void fraction and a decrease in pressure wave velocity, as shown in Figures 11 and 12. Within the range of low gas influx rate, the wave velocity decreases significantly. It is because of this that the compressibility of the gas increases remarkably, and the medium appears to be of high elasticity, though the density of gas-drilling mud two-phase flow changes slightly. With the increase in the gas influx rate and corresponding increase in the void fraction in the annulus, the compressibility of the two-phase unceasingly increases, which promotes the momentum and energy exchange in the interface. So, the pressure wave continuously decreases. When the void fraction increases to some extent following the increase in the gas influx rate, the decrease in wave velocity in the liquid slug is gradually less than the increase in wave velocity in the gas slug; thus, the decrease of wave velocity is slowed down for the growth of gas slug. Especially in the wellhead, a slight increase in wave velocity is observed.

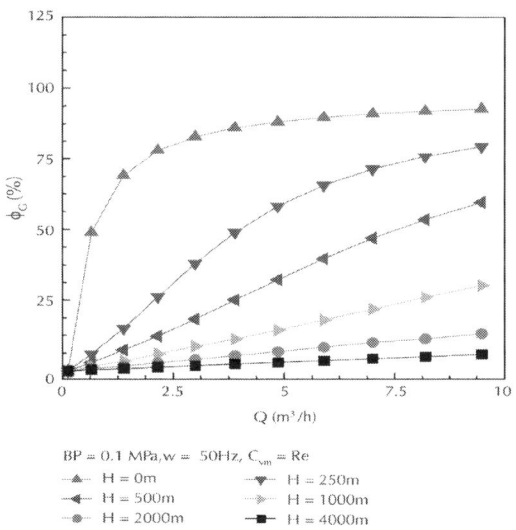

**Figure 11:** Effect of gas influx rate on the void fraction.

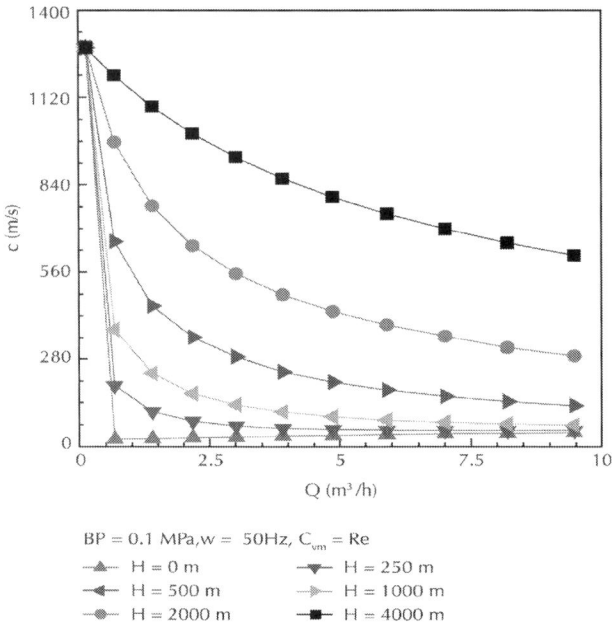

**Figure 12:** Effect of gas influx rate on the pressure wave velocity.

## Effect of Virtual Mass Force on Wave Velocity

Figure 13 illustrates the effect of virtual mass force on the wave velocity with the parameter of gas influx rate. As shown in Figure 15, under the same gas influx rate (0.594 m³/h, 1.098 m³/h, and 1.768 m³/h, resp.) the wave velocity curve of $C_{vm}$=Re diverges from the curve of $C_{vm}$=0 at the position close to the wellhead, whereas the wave velocity curve of the two types is almost coincided below the position of H=500 m. This divergence between the two types of curves is connected with the fact that significance of influence of interphase virtual mass forces increases together with the increase of the relative motion in the interphase. It is evident that the bottomhole pressure is hundreds of times higher than the wellhead pressure. As a result, the void fraction at the position close to the wellhead increases sharply, meanwhile the interphase momentum and energy exchanges are promoted. The virtual mass force can be described as the transfer of momentum between the gas phase and drilling mud phase caused by the relative motion in the interphase. If the relative motion in the interphase is quite weak, the value of virtual mass force will intend to approach 0, and the influence on the wave velocity can be ignored. However, if the relative motion is rather intense, the effect of virtual mass force on the wave velocity should not be ignored. Furthermore, taking the virtual mass force into account, the dispersion characteristic of the pressure wave weakens obviously. Compared with the pressure wave velocity calculated by ignoring the effect of virtual mass force, the calculated pressure wave velocity with a consideration of the virtual mass force is lower. Therefore, it is necessary to analyze the effect of the virtual mass force on the wave velocity at the position close to the wellhead in MPD operations.

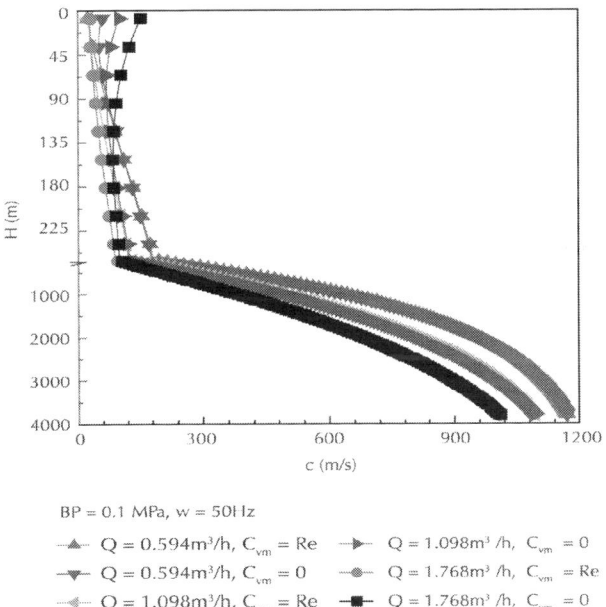

**Figure 13**: Effect of virtual mass force on the pressure wave velocity.

# Effect of Disturbance Angular Frequency on Wave Velocity

Figure 14 presents the pressure wave velocity in the annulus at different angular frequencies. According to Figures 5 and 9, the void fraction first slightly increases in the bottomhole and then sharply increases along the flow direction at the position close to the wellhead. At a fixed angular frequency, this phenomenon results in an overall decrease in the pressure wave velocity except for the section near the wellhead. At the section close to the wellhead, an opposite change trend of pressure wave velocity is observed for the transition of flow pattern from bubbly flow to slug flow due to the continuous increase in void fraction along the flow direction. In addition, it can be clearly seen from the curves that the wave velocity increases accompanying with the increase in the angular frequency above the position of H=500m. This property is not very distinct at the position below H=500 m for the low void fraction.

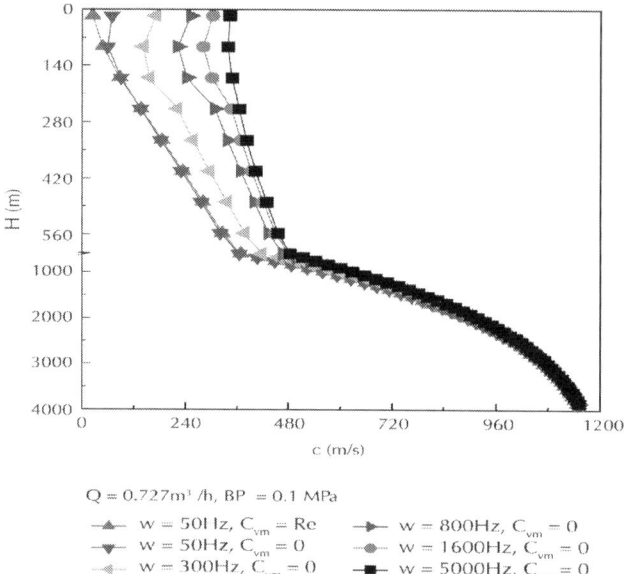

**Figure 14:** Wave velocity variations at different angular frequencies.

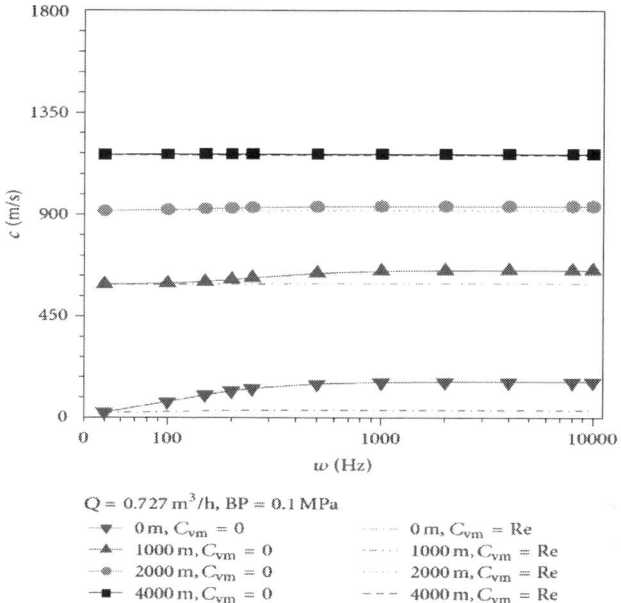

**Figure 15:** Effect of angular frequency on wave velocity.

Figure 15 shows the effect of frequency on the wave velocity in the gas-drilling mud flow ($C_{vm}=0$). The curves of the wave velocity of different positions reveal that the propagation velocity of pressure disturbances increases together with the growth of the angular frequency ($0<w<500Hz$). It proves that the pressure wave has an obvious dispersion characteristic in the two-phase flow. As the angular frequency increases in the range of less than 500 Hz, there is sufficient time to carry out the exchange of energy and momentum between two phases. It achieves a state of mechanical and thermodynamic equilibrium between the two phases. So the wave velocity increases gradually with the increase in the angular frequency at different depths of the annulus. It is considered that the wave velocity is mainly affected by the interphase mechanical and thermodynamic equilibrium at low anglular frequencies.w When the angular frequency reaches the value of w=500Hz, the pressure wave velocity achieves a constant value and remains on this level regardless of the further growth in angular frequency .

With the increase in angular frequency, there is not enough time for energy and momentum exchange between the gas-drilling mud two phases to reach the mechanical and thermodynamic equilibrium state, and thus the wave dispersion does not exist. At a high angular frequency, the wave velocity is mainly dominated by the mechanical and thermal nonequilibrium in the flow and keeps almost unchanged when the angular frequency is further increased. This is consistent with the influences of the angular frequency in the horizontal pipe [28]. Moreover, the effect of virtual mass force is also shown in this figure ($C_{vm}$=Re). It is observed that the wave velocity is significantly reduced when the virtual mass force is taken into consideration in the calculation of the wave velocity at the position close to the wellhead such as H=0m. It can be explained by the reduction of dispersion characteristic of a pressure wave. Along the wellbore from the wellhead to the bottomhole, the distinction gradually decreases.

## CONCLUSIONS

With full consideration of the important factors such as virtual mass force, drag force, gas void fraction, pressure, temperature, and angular frequency, a united wave velocity model has been proposed based on pressure drop gradient equations in MPD operations, gas-liquid two-fluid model, the gas-drilling mud equations of state, and small perturbation theory. Solved by the fourth-order explicit Runge-Kutta method, the model is used to predict wave velocity for different back pressures and gas influx rates in MPD operations. The main conclusions can be summarized as follows.

- With the introduction of virtual mass force and drag force, the united model agrees well with the previous experimental data. The united model can be used to accurately calculate the wave velocity in the annulus. The application of the model will be beneficial to further study the wave velocity at different gas influx rates and back pressures in MPD operations, reduce nonproductive times, and provide a reference for the drilling operations in extremely narrow pore/fracture windows existing conditions.

- The wave velocity and void fraction have different values with respect to well depth. In the annulus, the drop of pressure causes an increase in the void fraction along the flow direction. The void fraction increases first slightly and then sharply. Correspondingly, the wave velocity first gradually decreases in th bubbly flow and preliminary slug flow, and then the wave velocity slightly increases accompanying with the increase in the relative length ratio of gas slug to the liquid slug for the continuous increase in void fraction at the position close to the wellhead. The minimum wave velocity appears in the long gas slug flow.

- When the back pressure in the MPD operations increases, the void fraction at different depths of the annulus is reduced gradually, while the wave velocity in the two-phase flow tends to increase. Moreover, the influence of back pressure on the void fraction is greater at a position close to the wellhead than that in the bottomhole because of the great decrease in the pressure from the bottomhole to the wellhead. Also, the effect of back pressure on void fraction and wave velocity is decreased in the high back pressure range.

- The wave velocity is sensitive to the void fraction, but the void fraction is dominated by gas influx rate and pressure in the annulus, especially the gas influx rate. Since the compressibility of the gas and drilling mud two-phase fluid is improved with the increase in the influx rate, the void fraction increases greatly, and the wave velocity decreases significantly within the low gas influx rate range. When the void fraction is increased to some extent following the increase in the gas influx rate, the decrease of wave velocity is slowed for the growth of gas slug. Especially at the wellhead, a slight increase in wave velocity is observed at a high gas influx rate for the sharp increase in void fraction.

- It is necessary to analyze the effect of the virtual mass force on the wave velocity in MPD operations. Especially, at the position close to the wellhead the effect of virtual mass force is more prominent for the intense phase interactions. Taking the virtual mass force into account, the dispersion characteristic of the pressure wave weakens obviously. Compared with the results calculated by ignoring the effect of virtual mass force, the calculated pressure wave velocity with a consideration of the virtual mass force is lower.

- The effect of angular frequency on the propagation velocity of pressure wave appears at a low angular frequency. The propagation velocity of pressure disturbances increases together with the growth of the angular frequency (($0<w<500Hz$)). When the angular frequency reaches the value of $w=500Hz$, the pressure wave velocity achieves a constant value and remains on this level regardless of the further growth in angular frequency w.

# ACKNOWLEDGMENTS

Research work was cofinanced by the National Natural Science Foundation of China (nos. 51074135 and 51274170). Without their support, this work would not have been possible.

# REFERENCES

1. Y. Bu, F. Li, Z. Wang, and J. Li, "Preliminary study of air injection in annuli to manage pressure during cementing," Procedia Engineering, vol. 18, pp. 329–334, 2011.
2. P. Vieira, F. Torres, R. A. Qamar, and G. E. Marin, "Downhole pressure uncertainties related to deep wells drilling are safely and precisely ascertained using automated MPD technology," in Proceedings of the North Africa Technical Conference and Exhibition, Cairo, Egypt, February 2012.
3. S. Saeed and R. Lovorn, "Automated drilling systems for MPD—the reality," in Proceedings of the IADC/SPE Drilling

conference and Exhibition, San Diego, Calif, USA, March 2012.

4. W. A. Bacon, Consideration of compressibility effects for applied-back-pressure dynamic well control response to a gas kick in managed pressure drilling operations [M.S. thesis], University of Texas, Arlington, Tex, USA, 2011.

5. J. Choe, C. Hans, and J. Wold, "A modified two-phase well-control model and its computer applications as a training and educational tool," SPE Computer Applications, vol. 2, pp. 14–20, 1997.

6. J. A. Tarvin, I. Walton, and P. Wand, "Analysis of a gas kick taken in a deep well drilled with oil-based mud," in Proceedings of the Annual Technical Conference and Exhibition of the Society of Petroleum Engineers, pp. 255–264, Dallas, Tex, USA, 1991.

7. W. Guo, F. Honghai, and L. Gang, "Design and calculation of a MPD model with constant bottom hole pressure," Petroleum Exploration and Development, vol. 38, no. 1, pp. 103–108, 2011.

8. J. Zhou and G. Nygaard, "Automatic model-based control scheme for stabilizing pressure during dual-gradient drilling," Journal of Process Control, vol. 21, no. 8, pp. 1138–1147, 2011.

9. A. Ochoa, O. Acevedo, L. Nieto, J. E. Lambarria, and H. Pérez, "Successful application of MPD (Managed Pressure Drilling) for prevention, control, and detection of borehole ballooning in tight gas reservoir in Cuervito Field, Mexico," in Proceedings of the Canadian Unconventional Resources Conference (CURC '11), pp. 172–185, November 2011.

10. P. Vieira, F. Torres, R. A. Qamar, and G. E. Marin, "Downhole pressure uncertainties related to deep wells drilling are safely and precisely ascertained using automated MPD technology," in Proceedings of the North Africa Technical Conference and Exhibition, Cairo, Egypt, February 2012.

11. R. Ashena and J. Moghadasi, "Bottom hole pressure estimation using evolved neural networks by real coded ant colony optimization and genetic algorithm," Journal of Petroleum Science and Engineering, vol. 77, no. 3-4, pp. 375–385, 2011.
12. S. Wang, C. Qiang, and K. Bo, "Fluctuating pressure calculation during the progress of trip in managed pressure drilling," Advanced Materials Research, vol. 468-471, pp. 1736–1742, 2012.
13. I. Zubizarreta, "Pore pressure evolution, core damage and tripping out schedulesf: a computational fluid dynamics approach," in Proceedings of the SPE/IADC Drilling Conference and Exhibition, , Amsterdam, The Netherlands, March 2013.
14. C. H. Whitson, "Cyclic shut-in eliminates liquid-loading in gas wells," in Proceedings of the SPE/EAGE European Unconventional Resources Conference and Exhibition, Vienna, Austria, March 2012.
15. G. M. de Oliveira, A. T. Franco, C. O. R. Negrao, A. L. Martins, and R. A. Silva, "Modeling and validation of pressure propagation in drilling fluids pumped into a closed well," Journal of Petroleum Science and Engineering, vol. 103, pp. 61–71, 2012.
16. Y. Sato and H. Kanki, "Formulas for compression wave and oscillating flow in circular pipe," Applied Acoustics, vol. 69, no. 1, pp. 1–11, 2008.
17. J. Xie, B. Yu, X. Zhang, Q. Shao, and X. Song, "Numerical simulation of gas-liquid-solid three-phase flow in deep wells," Advances in Mechanical Engineering, vol. 2013, Article ID 951298, 10 pages, 2013.
18. W. Bacon, A. Tong, O. Gabaldon, C. Sugden, and P. V. Suryanarayana, "An improved dynamic well control response to a gas influx in managed pressure drilling operations," in Proceedings of the IADC/SPE Drilling Conference and Exhibition, San Diego, California, USA, March 2012.

19. J. I. Hage and D. ter Avest, "Borehole acoustics applied to kick detection," Journal of Petroleum Science and Engineering, vol. 12, no. 2, pp. 157–166, 1994.
20. A. M. R. Hopkins, Fluid dynamics and surface pressure fluctuations of two-dimensional turbulent boundary layers over densely distributed surface roughness [Ph.D. thesis], Virginia Polytechnic Institute and State University, Blacksburg, Va, USA, 2010.
21. Z. S. Fu, "The gas cut detect technology during drilling in Soviet Union," Petroleum Drilling Techniques, vol. 18, no. 1, pp. 19–21, 1990.
22. X. F. Li, C. X. Guan, and X. X. Sui, "The theory of gas influx detection of pressure wave and its application," Acta Petrolei Sinica, vol. 18, no. 3, pp. 128–133, 1997.
23. H. Li, Y. Meng, G. Li, et al., "Propagation of measurement-while-drilling mud pulse during high temperature deep well drilling operations," Mathematical Problems in Engineering, vol. 2013, Article ID 243670, 12 pages, 2013.
24. A. E. Ruggles, R. T. Lahey Jr., D. A. Drew, and H. A. Scarton, "Investigation of the propagation of pressure perturbations in bubbly air/water flows," Journal of Heat Transfer, vol. 110, no. 2, pp. 494–499, 1988.
25. H. J. W. M. Legius, H. E. A. Van Den Akker, and T. Narumo, "Measurements on wave propagation and bubble and slug velocities in cocurrent upward two-phase flow," Experimental Thermal and Fluid Science, vol. 15, no. 3, pp. 267–278, 1997.
26. K. Miyazaki and I. Nakajima, "Propagation of pressure wave in NItrogen-Mercury two-phase system," Journal of Nuclear Science and Technology, vol. 9, no. 5, pp. 315–318, 1972.
27. B. Bai, L. Guo, and X. Chen, "Pressure fluctuation for air-water two-phase flow," Journal of Hydrodynamics, vol. 18, no. 4, pp. 476–482, 2003.
28. F. Huang, M. Takahashi, and L. Guo, "Pressure wave propagation in air-water bubbly and slug flow," Progress in Nuclear Energy, vol. 47, no. 1–4, pp. 648–655, 2005.

29. R. F. Tangren, C. H. Dodge, and H. S. Seifert, "Compressibility effects in two-phase flow," Journal of Applied Physics, vol. 20, no. 7, pp. 637–645, 1949.
30. G. B. Wallis, One Dimensional Two Phase Flow, McGraw Hill, New York, NY, USA, 1969.
31. D. L. Nguyen, E. R. F. Winter, and M. Greiner, "Sonic velocity in two-phase systems," International Journal of Multiphase Flow, vol. 7, no. 3, pp. 311–320, 1981.
32. R. C. Mecredy and L. J. Hamilton, "The effects of nonequilibrium heat, mass and momentum transfer on two-phase sound speed," International Journal of Heat and Mass Transfer, vol. 15, no. 1, pp. 61–72, 1972.
33. K. H. Ardron and R. B. Duffey, "Acoustic wave propagation in a flowing liquid-vapour mixture," International Journal of Multiphase Flow, vol. 4, no. 3, pp. 303–322, 1978.
34. J. Xu and T. Chen, "Acoustic wave prediction in flowing steam-water two-phase mixture," International Journal of Heat and Mass Transfer, vol. 43, no. 7, pp. 1079–1088, 2000.
35. N. Chung, W. Lin, B. Pei, and Y. Hsu, "Model for sound velocity in a two-phase air-water bubbly flow," Nuclear Technology, vol. 99, no. 1, pp. 80–89, 1992.
36. X. Xu and J. Gong, "A united model for predicting pressure wave speeds in oil and gas two-phase pipeflows," Journal of Petroleum Science and Engineering, vol. 60, no. 3-4, pp. 150–160, 2008.
37. W. Kuczynski, "Characterization of pressure-wave propagation during the condensation of R404A and R134a refrigerants in pipe mini-channels that undergo periodic hydrodynamic disturbance," International Journal of Heat and Fluid Flow, vol. 40, pp. 135–150, 2013.
38. W. Kuczynski, "Modeling of the propagation of a pressure wave during the condensation process of R134a refrigerant in a pipe minichannel under the periodic conditions of hydrodynamic disturbances," International Journal of Heat and Mass Transfer, vol. 56, pp. 715–723, 2013.

39. Y. Li, C. Li, E. Chen, and Y. Ying, "Pressure wave propagation characteristics in a two-phase flow pipeline for liquid-propellant rocket," Aerospace Science and Technology, vol. 15, no. 6, pp. 453–464, 2011.
40. T. Kanagawa, M. Watanabe, T. Yano, and S. Fujikawa, "Nonlinear wave equations for pressure wave propagation in liquids containing gas bubbles (comparison between two-fluid model and mixture model)," Journal of Fluid Science and Technology, vol. 6, no. 6, pp. 836–849, 2011.
41. E. Chen, Y. Li, and Y. Ying, "Numerical investigation on pressure wave propagation speed of gas-liquid two-phase flow in pump pipeline," Journal of Aerospace Power, vol. 25, no. 4, pp. 754–760, 2010.
42. F. Li and L. He, "Investigation on pressure wave propagation characteristics in slug flow field with increasing gas flow rate," Petroleum Geology and Recovery Efficiency, vol. 11, no. 5, pp. 78–80, 2004.
43. C. S. Martin and M. Padmanabhan, "Pressure pulse propagation in two-component slug flow," Journal of Fluids Engineering, vol. 101, no. 1, pp. 44–52, 1979
44. R. E. Henry, "Pressure wave propagation in two-phase mixtures," Chemical Engineering Progress, vol. 66, no. 1, pp. 1–10, 1969.
45. X. Liu, B. Li, and Y. Yue, "Transmission behavior of mud-pressure pulse along well bore," Journal of Hydrodynamics B, vol. 19, no. 2, pp. 236–240, 2007.
46. X. Wang and J. Zhang, "The research of pressure wave Pulsation in mud pulse transmitting," Journal of Chongqing University of Science and Technology, vol. 14, no. 2, pp. 55–58, 2012.
47. J.-W. Park, D. A. Drew, and R. T. Lahey Jr., "The analysis of void wave propagation in adiabatic monodispersed bubbly two-phase flows using an ensemble-averaged two-fluid model," International Journal of Multiphase Flow, vol. 24, no. 7, pp. 1205–1244, 1998.

48. G. Arnold, Entropy and objectivity as constraints upons constitutive equations for two-fluid modeling of multiphase flow [Ph.D. thesis], Rensselaer Polytechnic Institute, New York, NY, USA, 1998.
49. P. M. Dranchuk and J. H. Abou-Kassem, "Calculation of Z factors natural gases using equitation of state," Journal of Canadian Petroleum Technology, vol. 14, no. 3, pp. 34–36, 1975.
50. L. Yarborough and K. R. Hall, "How to solve equation of state for z-factors," Oil and Gas Journal, vol. 72, no. 7, pp. 86–88, 1974.
51. A. R. Hasan and C. S. Kabir, "Wellbore heat-transfer modeling and applications," Journal of Petroleum Science and Engineering, vol. 86, pp. 127–136, 2012.
52. H. Zhu, Y. Lin, D. Zeng, D. Zhang, and F. Wang, "Calculation analysis of sustained casing pressure in gas wells," Petroleum Science, vol. 9, no. 1, pp. 66–74, 2012.
53. J. Orkiszewski, "Predicting two-phase pressure drops in vertical pipe," Journal of Petroleum Technology, vol. 6, no. 6, pp. 829–838, 1967.
54. M. Ishii and K. Mishima, "Two-fluid model and hydrodynamic constitutive relations," Nuclear Engineering and Design, vol. 82, no. 2-3, pp. 107–126, 1984.
55. M. J. Sanchez, Comparison of correlations for predicting pressure losses in vertical multiphase annular flow [M.S. thesis], University of Tulsa, Tulsa, Okla, USA, 1972.

# Chapter 2

# Diffusion of Chemically Reactive Species in Casson Fluid Flow over an Unsteady Stretching Surface in Porous Medium in the Presence of a Magnetic Field

Gilbert Makanda, Sachin Shaw, and Precious Sibanda

School of Mathematics, Statistics and Computer Science, University of KwaZulu-Natal, Private Bag X01, Scottsville, Pietermaritzburg 3209, South Africa

## ABSTRACT

A study is performed on two-dimensional flow and diffusion of chemically reactive species of Casson fluid from an unsteady stretching surface in porous medium in the presence of a magnetic field. The boundary layer velocity, temperature, and concentration profiles are numerically computed for different governing parameters. The paper intends to show unique results of a combination of heat transfer and chemical reaction in Casson fluid flow. The resulting partial differential equations are converted to a system of ordinary differential equations using the appropriate similarity transformation, which are solved by using the Runge-Kutta-Fehlberg numerical scheme. The results in this work are validated by the comparison with other authors.

## INTRODUCTION

The study of Casson fluid has attracted attention to many researchers due to its application in the field of metallurgy, food processing, drilling operations, and bioengineering operations. Its application extends to the manufacturing of pharmaceutical products, coal in water, china clay, paints, synthetic lubricants, and biological fluids such as synovial fluids, sewage sludge, jelly, tomato sauce, honey, soup, and blood due to its contents such as plasma, fibrinogen, and protein, making the study of Casson fluid important in fluid dynamics. Casson fluid is classified as a non-Newtonian fluid due to its rheological characteristics. These characteristics show shear stress-strain relationships that are significantly different from Newtonian fluids. The study of non-Newtonian fluids has not been thoroughly covered due to the complex representation of their constitutive equations. It is therefore important to undertake this study of Casson fluid. Most studies have concentrated on viscoelastic fluids in which different constitutive equations have been suggested. This work can be applied to chemical processing equipment in which some fluids react chemically with some ingredients present in them.

The driving force for mass transfer is a combination of temperature and concentration gradients. In this study the effect of chemical reaction on the fluid is considered as in Mukhopadhyay and Vajravelu [1]. The study of boundary layer flow over a stretching sheet has been studied by Mukhopadhyay et al. [2], among others, who investigated Casson fluid flow over an unsteady stretching surface. In their work they did not consider mass transfer and they considered a different wall temperature expression. Abd El-Aziz [3] studied mixed convection flow of a micropolar fluid from an unsteady stretching surface with viscous dissipation. In this work he considered a similar stretching velocity, wall temperature, and wall concentration distribution. We extended the work of Grubka and Bobba [4] who investigated heat transfer characteristics of a continuous stretching surface with variable temperature in which we introduced the MHD and porous medium source terms and chemical reaction effects. Sharidan et al. [5] studied similarity solutions for the unsteady boundary layer flow and heat transfer due to a stretching sheet; Nadeem et al. [6, 7] and Ahmed and Nazar [8] also studied Casson fluid over a stretching sheet and in their paper they assumed that the velocity of the stretching surface is linearly proportional to the distance from fixed origin.

In the study of non-Newtonian fluids many authors have studied the flow of blood as Casson fluid. The studies were carried out by Rohlf and Tenti [9], among others, who investigated the role of Womersley number in pulsatile blood flow, a theoretical study of the Casson model; Sankar and Lee [10, 11] investigated two-fluid nonlinear model for flow in catheterized blood vessels and two-fluid Casson model for pulsatile blood flow through stenosed arteries, respectively. Shaw et al. [12] studied pulsatile Casson fluid flow through stenosed bifurcated artery. In relation to blood flow there are other research works that were done in different geometries such as flows in microslit channels, slightly curved channels, and peristaltic transport as in [13–15].

The study of Casson fluid in porous media was also studied by Nadeem et al. [16] who considered MHD three-dimensional Casson fluid flow past a porous linearly stretching sheet. Dash et al.

[17] studied Casson fluid flow in a pipe filled with homogeneous porous medium. Tripathi [18] investigated the transient peristaltic heat flow through a finite porous channel. Pramanik [19] studied Casson fluid flow and heat transfer past an exponentially porous stretching surface in the presence of thermal radiation. Ramachandra et al. [20] investigated flow and heat transfer of Casson fluid from a horizontal circular cylinder with partial slip in a non-Darcy porous medium. In their work they considered slip conditions on the wall.

There are many studies that investigated fluid flow with chemical reactions. Kameswaran et al. [21] investigated homogeneous-heterogeneous reactions in a nanofluid flow due to a porous stretching sheet, Shaw et al. [22] studied homogeneous-heterogeneous reactions in a nanofluid flow due to a porous stretching sheet, and Chamkha et al. [23] investigated similarity solutions for unsteady heat and mass transfer from a stretching surface embedded in a porous medium with suction/injection and chemical reaction effects. Although there are many applications and use of non-Newtonian fluids in industry, the study of Casson fluid has not been thoroughly investigated for heat and mass transfer past a stretching surface. In this work we extend the work of Mukhopadhyay and Vajravelu [1] and Grubka and Bobba [4] in which the energy equation, the source terms for porous medium, and magnetic field are introduced. Similarity transformations are used to convert the partial differential equations into ordinary differential equations which are then solved by using Runge-Kutta-Fehlberg integration scheme and the successive linearization method described by Makanda et al. [24]. In this work we investigate the effect of varying unsteadiness parameter, Casson, Schmidt, and Prandtl numbers, and the reaction rate parameter on the velocity, temperature, and concentration profiles with the depiction of graphical illustrations.

# MATHEMATICAL FORMULATION

Consider two-dimensional laminar boundary layer flow, temperature, and mass transfer of an incompressible Casson fluid

# Diffusion of Chemically Reactive Species in Casson Fluid Flow...

flow over an unsteady stretching sheet. The flow of heat and mass transfer starts at $t = 0$. The sheet is pulled out of the slit at the origin $(x = 0, y = 0)$ and moves with velocity $U(x, t) = ax/(1 - \alpha t)$, $a > 0$, $\alpha \geq 0$ are constants, and $a$ is the initial stretching rate. The rheological equation of state for an isotropic and incompressible flow of a Casson fluid is given as in [1, 2, 20] by

$$\tau_{ij} = \begin{cases} 2\left(\mu_B + \dfrac{P_y}{\sqrt{2\pi}}\right) e_{ij}, & \pi > \pi_c \\ 2\left(\mu_B + \dfrac{P_y}{\sqrt{2\pi_c}}\right) e_{ij}, & \pi < \pi_c, \end{cases} \quad (1)$$

where $\pi = e_{ij} e_{ij}$ and $e_{ij}$ is the $(i, j)$th component of the deformation rate, $\pi$ is the product of the deformation rate with itself, $\pi_c$ is a critical value of this product based on the non-Newtonian model, $\mu B$ is the plastic dynamic viscosity of the non-Newtonian fluid, and $P_y$ is the yield stress of the fluid. Given that $T_w$ and $C_w$ are, respectively, the temperature and concentration at the sheet and $T_\infty$ and $C_\infty$ are, respectively, the ambient conditions, the positive $x$ coordinate is measured along the stretching sheet and the positive $y$ coordinate is measured perpendicular to the sheet. It is assumed that both temperature and concentration at the surface vary with distance from the origin and time. The temperature $T_w$ and concentration $C_w$ at the surface are therefore given by

$$T_w(x,t) = T_\infty + \frac{bx}{(1-\alpha t)^2}, \quad C_w(x,t) = C_\infty + \frac{cx}{(1-\alpha t)^2}, \quad (2)$$

Where $b$ and $c$ are constants. The surface temperature and surface concentration increase if $b$ and $c$ are positive and reduce if they are negative from $T_\infty$ and $C_\infty$ at the origin to $x$ and the temperature and concentration increase/decrease along the sheet. It is assumed that radiation effects and viscous dissipation are negligible. The expressions $U_w(x, t)$, $T_w(x, t)$, and $C_w(x, t)$ are only valid for $t <$

$\alpha^{-1}$ but not when $\alpha = 0$. Under these assumptions the governing equations in this flow are given as

$$\frac{\partial}{\partial x}(u) + \frac{\partial}{\partial y}(v) = 0,$$

$$\frac{\partial u}{\partial t} + u\frac{\partial u}{\partial x} + v\frac{\partial u}{\partial y} = \nu\left(1 + \frac{1}{\beta}\right)\frac{\partial^2 u}{\partial y^2} - \frac{\nu}{K}u - \frac{\sigma B_0^2}{\rho}u, \quad (3)$$

$$\frac{\partial T}{\partial t} + u\frac{\partial T}{\partial x} + v\frac{\partial T}{\partial y} = \alpha_0 \frac{\partial^2 T}{\partial y^2},$$

$$\frac{\partial C}{\partial t} + u\frac{\partial C}{\partial x} + v\frac{\partial C}{\partial y} = D\frac{\partial^2 C}{\partial y^2} - k(C - C_\infty), \quad (4)$$

where $\nu$ is kinematic viscosity of Casson fluid, $\beta = \mu_B \sqrt{2\pi_c}/P_y$ is the non-Newtonian Casson parameter, $K$ is the permeability of the porous medium, $\sigma$ is the electrical conductivity, $B_0$ is the strength of the magnetic field, $\rho$ is the density of the Casson fluid, $D$ is the diffusion coefficient of species in the fluid, $\alpha_0$ is the thermal diffusivity, and $k(t) = k_0/(1 - \alpha t)$ is the time dependent reaction rate, where $k > 0$ represents destructive reaction, $k < 0$ represents constructive reaction, and $k_0$ is a constant. The boundary conditions are given as

$$u = U(x,t), \quad v = 0, \quad T = T_w(x,t),$$

$$C = C_w(x,t), \quad y = 0,$$

$$u \longrightarrow 0, \quad T \longrightarrow T_\infty, \quad C \longrightarrow C_\infty, \quad \text{as } y \longrightarrow \infty, \quad (5)$$

where the subscript $\infty$ refers to the ambient condition.

We introduce the nondimensional variables

$$u = \frac{\partial \psi}{\partial y}, \quad v = -\frac{\partial \psi}{\partial x},$$

$$\eta = \sqrt{\frac{a}{\nu(1-\alpha t)}}\, y, \quad \psi = \sqrt{\frac{\nu a}{1-\alpha t}}\, x f(\eta),$$

$$T_w(x,t) = T_\infty + \frac{bx}{(1-\alpha t)^2}\theta(\eta),$$

$$C_w(x,t) = C_\infty + \frac{cx}{(1-\alpha t)^2}\phi(\eta), \tag{6}$$

where $\psi(x, y, t)$ is the stream function which satisfies the continuity equation (3). The velocity components are defined as

$$u = \frac{\partial \psi}{\partial y} = U_w f'(\eta), \quad v = -\frac{\partial \psi}{\partial x} = -\sqrt{\frac{\nu a}{1-\alpha t}}. \tag{7}$$

The governing equations reduce to

$$\left(1+\frac{1}{\beta}\right)f''' + ff'' - f'^2 - \frac{A}{2}\left(2f' + \eta f''\right)$$
$$- \left(\Lambda + M^2\right)f' = 0,$$

$$\frac{1}{\Pr}\theta'' + f\theta' - f'\theta - \frac{A}{2}\left(4\theta + \eta\theta'\right) = 0,$$

$$\frac{1}{Sc}\phi'' + f\phi' - f'\phi - \frac{A}{2}\left(4\phi + \eta\phi'\right) - R\phi = 0, \tag{8}$$

with boundary conditions

$$f(0) = 0, \quad f'(0) = 1, \quad \theta(0) = 1, \quad \phi(0) = 1,$$

$$f'(\infty) \longrightarrow 0, \quad \theta(\infty) \longrightarrow 0, \quad \phi(\infty) \longrightarrow 0, \quad (9)$$

where $A = \alpha/a$ is the unsteadiness parameter, $\Pr = v/\alpha\, 0$ is the Prandtl number, $Sc\ v = ]/D$ is the Schmidt number, and $R = k_0/a$ is the reaction parameter. $\Lambda = v(1-\alpha t)/aK$ is the permeability parameter; $M^2 = \sigma B_0^2 (1-\alpha t)/\rho a$ is the magnetic parameter. The nondimensional temperature and concentration are, respectively, given by $\theta = (T-T_\infty)/(T_w - T_\infty)$ and $\phi = (C-C_\infty)/(Cw - C_\infty)$.

The parameters of engineering interests are the local skin friction and the Nusselt and Sherwood numbers which are defined as

$$C_{fx} = 2\left(1 + \frac{1}{\beta}\right)\operatorname{Re}_x^{-1/2} f''(0). \quad (10)$$

The local Nusselt and Sherwood numbers are defined as

$$\mathrm{Nu}_x = \frac{x}{\alpha_0} \frac{q_w}{(T_w - T_\infty)}, \quad \mathrm{Sh}_x = \frac{x}{D} \frac{J_w}{(C_w - C_\infty)}, \quad (11)$$

$$q_w = -\alpha_0 \left[\frac{\partial T}{\partial y}\right]_{y=0}, \quad J_w = -D\left[\frac{\partial C}{\partial y}\right]_{y=0}. \quad (12)$$

Using expressions (11) and (6),

$$\mathrm{Nu}_x = -\operatorname{Re}_x^{1/2} \theta'(0), \quad \mathrm{Sh}_x = -\operatorname{Re}_x^{1/2} \phi'(0), \quad (13)$$

where $\mathrm{Re}_x$ is the Reynolds number defined as $\mathrm{Re}_x = U_w x/v$. It is important at this stage to mention that the system of equations

(8)-(9) reduce to those of Grubka and Bobba [4] when $(1/\beta \to 0)$, $A = \lambda = M = Sc = 0$. The present problem reduces to that of Grubka and Bobba [4], $A = 0$ denote steady flow, and in their paper they obtained an exact solution in terms of Kummer's functions written in terms of the confluent hypergeometric functions.

The solution of the boundary value problem (8)-(9) was solved using the Runge-Kutta-Fehlberg integration scheme. In the method we choose finite values of $\eta \to \infty$. This value is the boundary layer thickness given by $\eta \infty$. We begin by choosing an initial guess of $\eta \infty$ to obtain the values $f''(0)$, $-\theta'(0)$, and $-\phi'(0)$. by $10^{-6}$.

# RESULTS AND DISCUSSION

To obtain a clear understanding of the flow of Casson fluid, we discuss the physics of the problem by studying the effects of the unsteadiness $(A)$, permeability $(\Lambda)$, magnetic $(M)$, Prandtl (Pr), Schmidt (Sc), and reaction rate $(R)$ numbers on velocity, temperature, and concentration profiles. We also study the variation of skin friction, the local Nusselt number, and the Sherwood number with unsteadiness parameter. For validation of the numerical method used in this study, results for the Nusselt number $-\theta(0)$ for the Newtonian fluid were compared to those of Abd El-Aziz [3] and Grubka and Bobba [4] for the unsteadiness parameter $A = \Lambda = M = Sc = 0$. The comparison is shown in Table 1 and it is found to be in agreement with at least four decimal places. To further verify the accuracy of the numerical scheme used, the successive linearization method (SLM) was used and there was agreement with Runge-Kutta-Fehlberg integration scheme.

**Table 1:** Comparison of the values of $\theta'-(0)$ for $\Lambda = M = Sc = 0$ and various values of $A$ and Pr with previously published data

| A | Pr | Grubka and Bobba [4] | Abd El-Aziz [3] | SLM | Present results |
|---|---|---|---|---|---|
| 0 | 0.72 | 0.8086 | 0.80873135 | 0.80873007 | 0.80863761 |
|   | 1 | 1.0000 | 1.00000000 | 1.00000000 | 1.00000006 |
|   | 3 | 1.9237 | 1.92368255 | 1.92367361 | 1.92367736 |
|   | 10 | 3.7207 | 3.72067395 | 3.72066225 | 3.72066701 |

To get a clear understanding of the behavior of velocity, temperature, and concentration profiles of Casson fluid, a detailed numerical calculation is done for different parameter values that describe the nature of flow and the results are depicted through Figures 1–5.

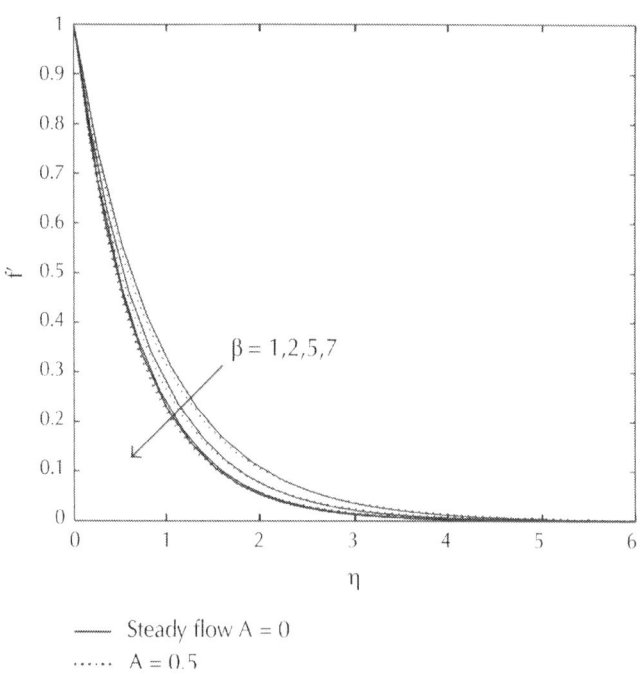

(a)

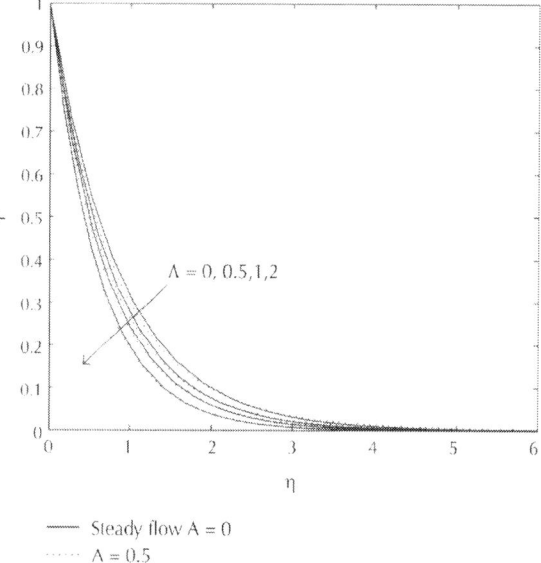

(b)

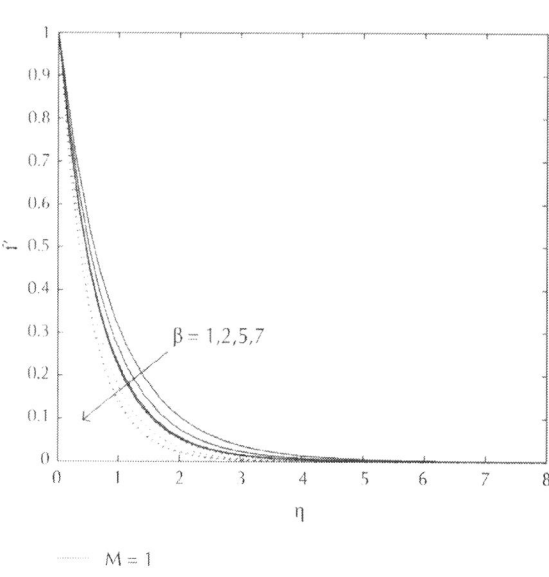

(c)

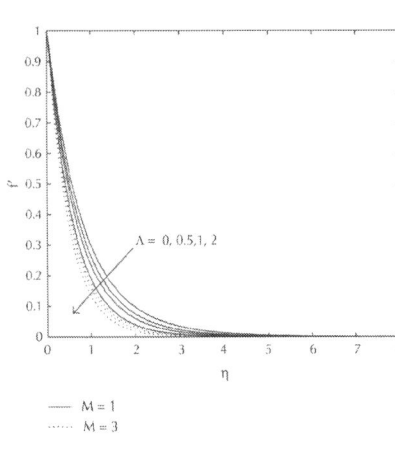

(d)

**Figure 1:** Effects of different parameters on velocity profiles. (a) Velocity profiles for different values of the Casson parameter $\beta$ and unsteadiness parameter $A$ at $\Lambda = 0.5, M = 1, \Pr = 1, Sc = 1,$ and $R = 0.5$ (b) Velocity profiles for different values of the permeability parameter $\Lambda$ and unsteadiness parameter $A$ at $\beta = 2, M = 1, \Pr = 1, Sc = 1,$ and $R = 0.5$ (c) Velocity profiles for different values of the Casson parameter $\beta$ and magnetic parameter $M$ at $A = 0.5, \Lambda = 0.5, \Pr = 1, Sc = 1,$ and $R = 0.5$ (d) Velocity profiles for different values of the permeability parameter $\Lambda$ and magnetic parameter $M$ at $\beta = 2, A = 0.5, \Pr = 1, Sc = 1,$ and $R = 0.5$.

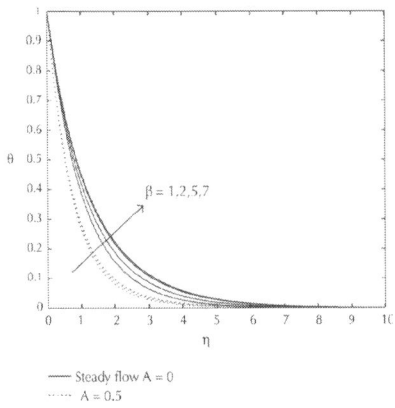

(a)

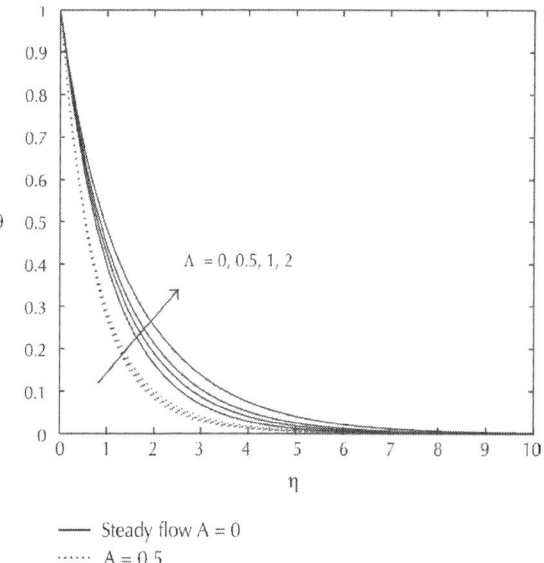

(b)

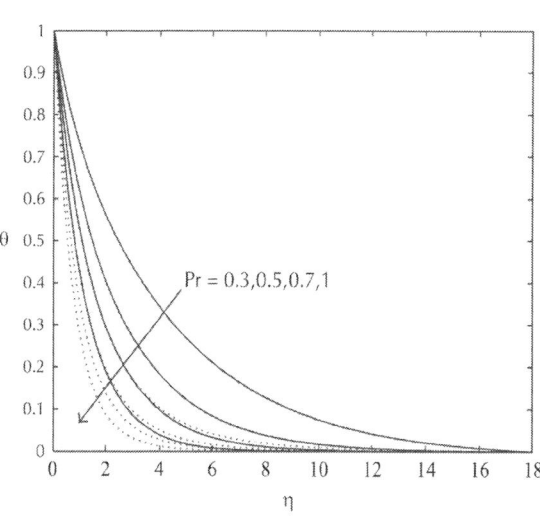

(c)

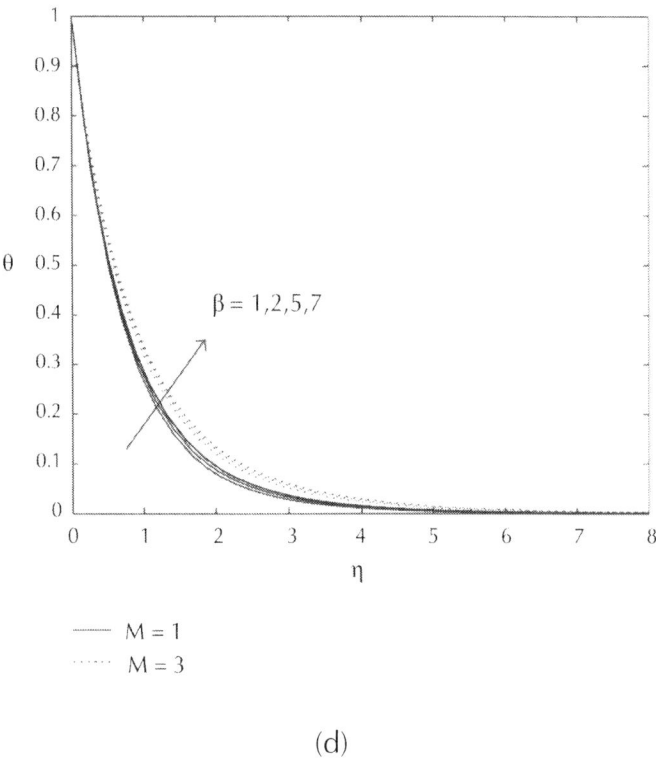

(d)

**Figure 2:** Effects of different parameters on temperature profiles. (a) Temperature profiles for different values of the Casson parameter $\beta$ and unsteadiness parameter $A$ at $\Lambda = 0.5, M = 1$, $Pr = 1$, $Sc = 1$, and $R = 0.5$(b) Temperature profiles for different values of the permeability parameter $\Lambda$ and unsteadiness parameter $A$ at $\beta = 2, M = 1$, $Pr = 1$, $Sc = 1$, and $R = 0.5$(c) Temperature profiles for different values of the Prandtl numbers $Pr$ and unsteadiness parameter $A$ at $\beta = 2$, $\Lambda = 0.5, M = 1$, $Sc = 1$, and $R = 0.5$(d) Temperature profiles for different values of the Casson parameter $\beta$ and magnetic parameter $M$ at $A = 0.5$, $\Lambda = 0.5$, $Pr = 1$, $Sc = 1$, and $R = 0.5$.

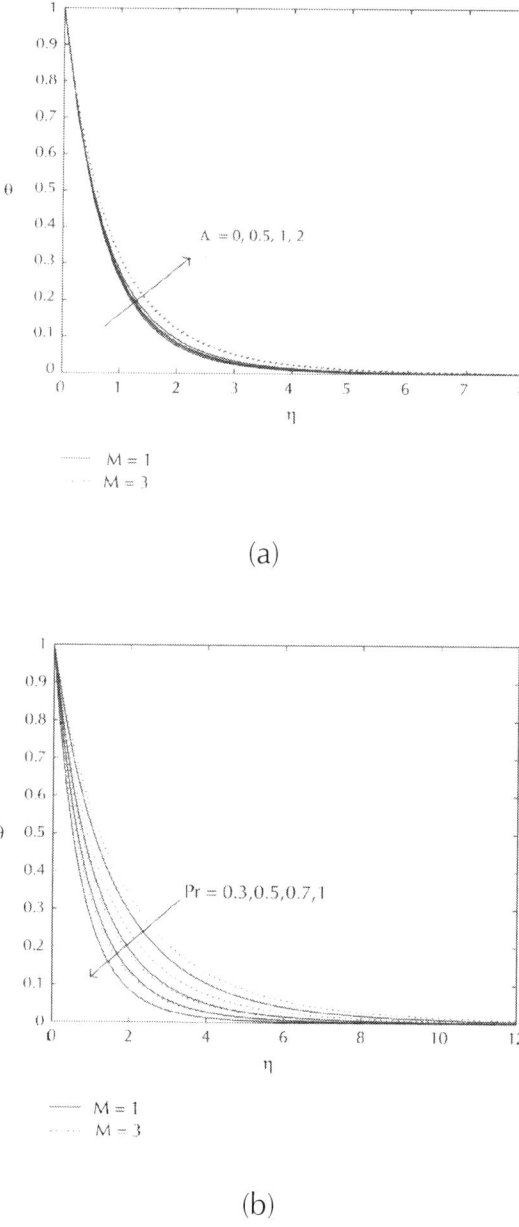

**Figure 3:** Effects of different parameters on temperature profiles. (a) Temperature profiles for different values of the permeability parameter $\Lambda$ and magnetic parameter $M$ at $\beta = 2$, $A = 0.5$, $Pr = 1$, Sc

= 1, and $R = 0.5$(b) Temperature profiles for different values of the Prandtl number Pr and magnetic parameter $M$ at $\beta = 2$, $A = 0.5$, $\Lambda = 0.5$, $Sc = 1$, and $R = 0.5$.

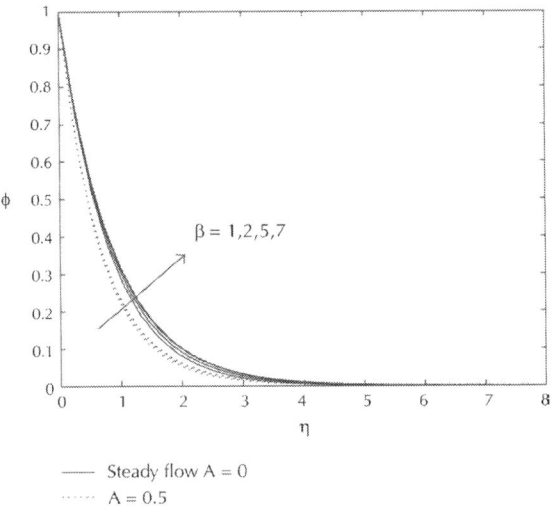

(a)

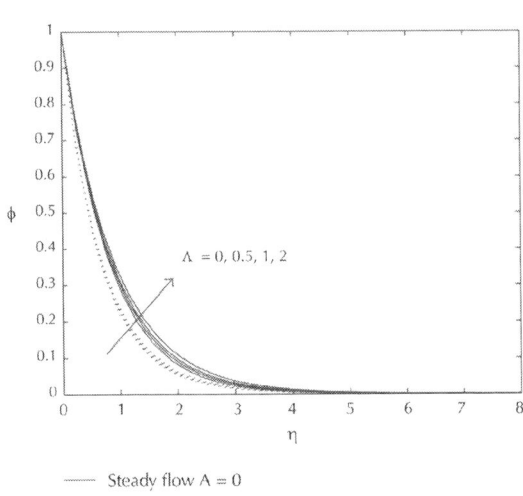

(b)

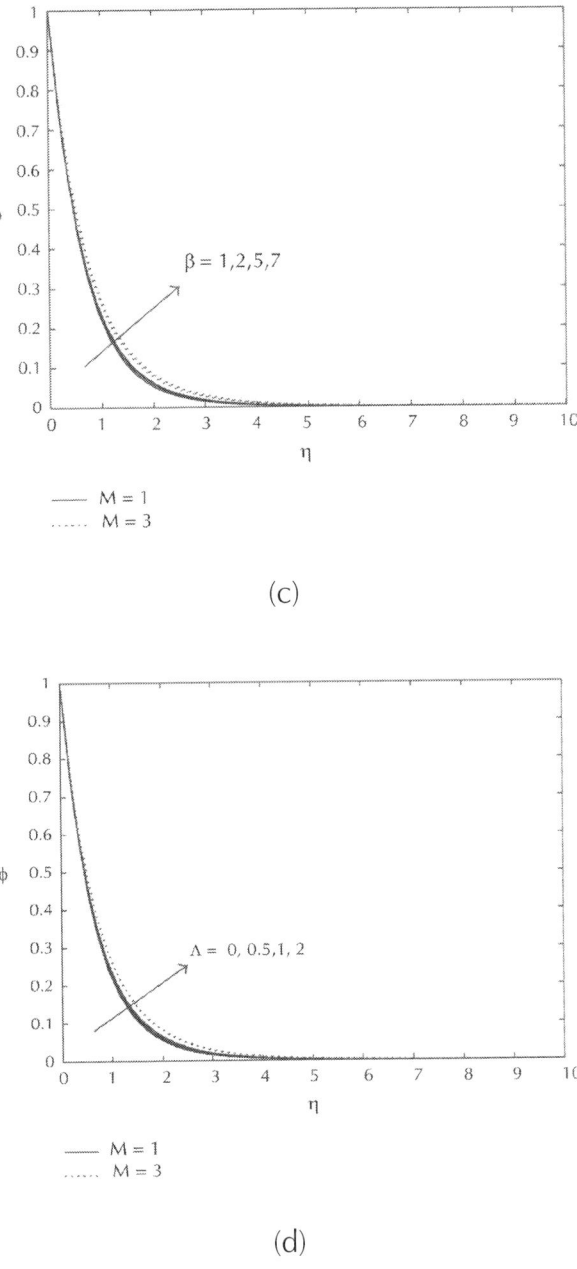

**Figure 4:** Effects of different parameters on concentration profiles. (a) Concentration profiles for different values of the Casson param-

eter $\beta$ and unsteadiness parameter $A$ at $\Lambda = 0.5, M= 1$, $\Pr = 1$, $Sc = 1$, and $R = 0.5$ (b) Concentration profiles for different values of the permeability parameter $\Lambda$ and unsteadiness parameter $A$ at $\beta = 2, M = 1$, $\Pr = 1$, $Sc = 1$, and $R = 0.5$ (c) Concentration profiles for different values of the Casson parameter $\beta$ and magnetic parameter $M$ at $A = 0.5$, $\Lambda = 0.5$, $\Pr = 1$, $Sc = 1$, and $R = 0.5$ (d) Concentration profiles for different values of the permeability parameter $\Lambda$ and magnetic parameter $M$ at $A = 0.5$, $\beta = 2$, $\Pr = 1$, $Sc = 1$, and $R = 0.5$.

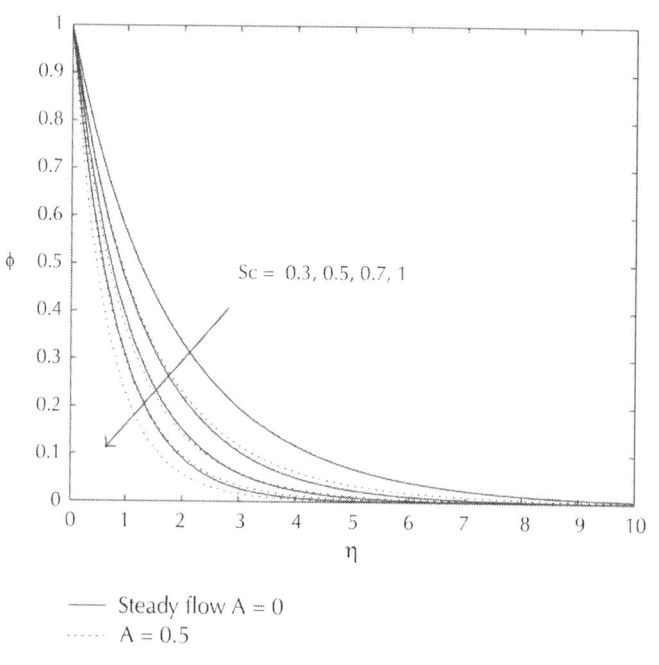

(a)

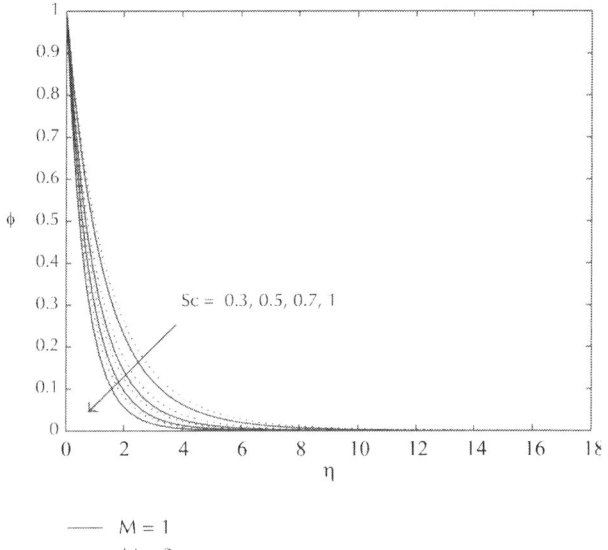

(b)

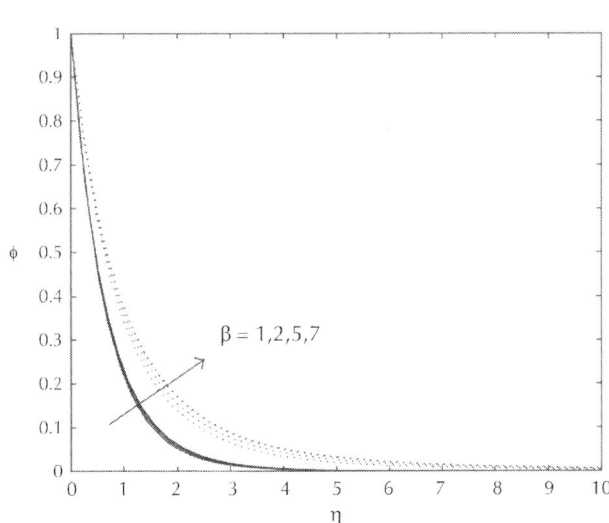

(c)

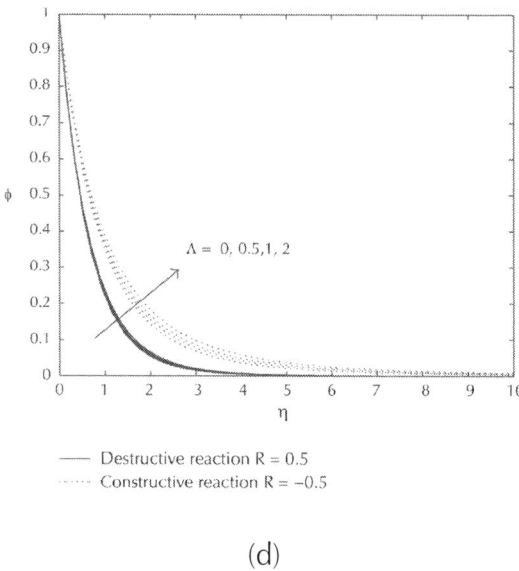

(d)

**Figure 5:** Effects of different parameters on concentration profiles. (a) Concentration profiles for different values of the Schmidt number Sc and unsteadiness parameter $A$ at $\beta = 2$, $\Lambda = 0.5$, $Pr = 1$, $M = 1$, and $R = 0.5$ (b) Concentration profiles for different values of the Schmidt number Sc and magnetic parameter $M$ at $A = 0.5$, $\beta = 2$, $\Lambda = 0.5$, $Pr = 1$, and $R = 0.5$ (c) Concentration profiles for different values of the Casson parameter $\beta$ and reaction rate $R$ at $A = 0.5$, $M = 1$, $\Lambda = 0.5$, $Pr = 1$, and $Sc = 1$ (d) Concentration profiles for different values of the permeability parameter $\Lambda$ and reaction rate $R$ at $A = 0.5$, $\beta = 2$, $M = 1$, $Pr = 1$, and $Sc = 1$.

Figure 1 shows that increasing the Casson parameter $\beta$ and the permeability parameter $\Lambda$ decreases velocity profiles. Both situations retard fluidmotion across the boundary layer as shown in Figures 1(a) to 1(d). The velocity profiles for unsteady flow become similar to steady flow far away from the boundary at $\eta > 2$. This indicates that for $A = 0.5$ the flow away from the boundary is steady. Increasing $A$ retards velocity profiles across the boundary layer as shown in Figures 1(a) and 1(b). In Figure 1(b) it is noted that increasing $A$ reduces velocity profiles and the reverse effect is noted around $\eta = 2.5$ indicating slightly steady flow far away from the boundary. A slightly different observation is noted if the magnetic parameter is increased from $M = 1$ to $M = 3$; the velocity profiles decrease significantly reducing themomentumboundary layer thickness as shown in Figures 1(c) and 1(d).

Figure 2 shows the variation of different parameters with temperature profiles. Increasing the Casson parameter increases temperature profiles; this decreases the yield stress. The fluid behaves as Newtonian as $\beta$ increases; this retards fluid motion. The effect of increasing $\beta$ leads to enhancing the temperature profiles for both steady and unsteady flows as shown in Figure 2(a). The same observation is noted in Figure 2(d). Increasing the permeability parameter $\Lambda$ enhances temperature profiles. High values of $\Lambda$ lead to the reduction of fluid velocity enhancing temperature profiles. This effect is more pronounced in steady flow and the thickening of the thermal boundary layer increases as $\Lambda$ increases. In Figure 2(c) increasing the Prandtl number Pr reduces temperature profiles; Pr is inversely proportional to the thermal diffusivity which is low for Casson fluid. In Figures 2(d), 3(a), and 3(b), increasing $M$ increases temperature profiles; the slowing down of fluid flow leads to the buildup of heat. It is more pronounced in low values of $M$. Increasing $A$ leads to the reduction of temperature profiles; less heat is transferred from the sheet to the fluid; hence, temperature decreases as depicted in Figures 2(a), 2(b), and 2(c).

In Figure 4, increasing the Casson parameter $\beta$ results in the increase of the solute concentration; as the fluid becomes Newtonian, solute transfer is enhanced as shown in Figures 4(a), 4(d), and 5(c). This is more pronounced in steady flow. The opposite effect is noted in increasing $A$ which reduces solute concentration. Increasing $\Lambda$ increases concentration profiles; since this reduces fluidmotion, it leads to the buildup of solute concentration as shown in Figures 4(b), 4(d), and 5(d). Increasing Sc decreases concentration profiles; the diffusion rate is smaller inCasson fluid and hence reducing solute concentration as depicted in Figures 5(a) and 5(b). In Figures 4(c), 4(d), and 5(b), increasing $M$ increases profiles; the slowing down of fluid flow leads to the buildup of mass. It is more pronounced in low values of $M$.

Figures 5(c) and 5(d) show the effects of reaction rate $R$ on the concentration profiles in Casson fluid in porous medium. Increasing the absolute value of $R$ decreases the concentration profiles. The reaction rate is in general a destructive agent which leads to the reduction of the solute boundary layer. It shows the increase of solute boundary layer in the case of constructive reaction $R < 0$ and the decrease of solute boundary layer in the case of destructive reaction $R > 0$. The effect of increasing $\beta$ and $\Lambda$ is more pronounced in unsteady flow. Figures 6(a) and 6(b) show effects of Casson parameter $\beta$ and permeability parameter $\Lambda$ on velocity gradient at the wall with unsteadiness parameter. The magnitude of $f''(0)$ related to skin friction decreases with increasing unsteadiness parameter $A$ and also with Casson parameter $\beta$ and permeability parameter $\Lambda$. The magnitude of the heat transfer rate at the surface $-\theta'(0)$ decreases with increasing $\beta$ and $\Lambda$ and increases with $A$ as shown in Figure 7. The mass transfer coefficient at the surface $-\phi'(0)$ decreases with increasing $\beta$ and $\Lambda$ and increases with $A$ as shown in Figure 8.

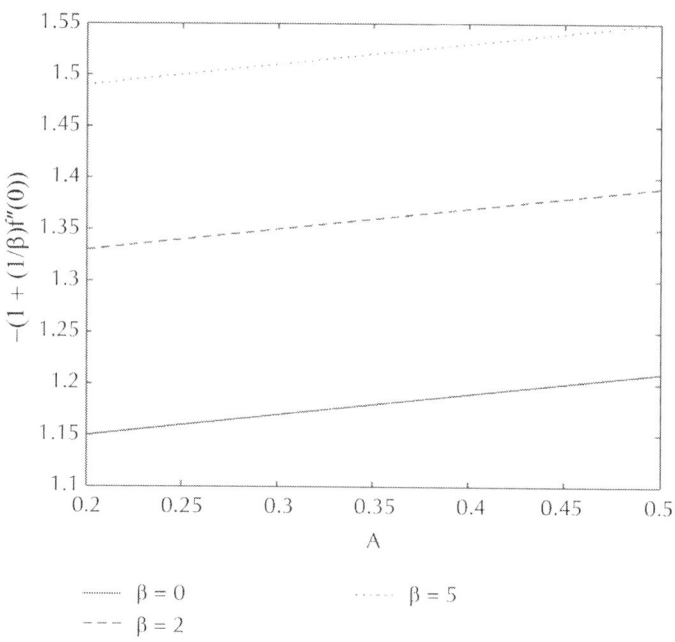

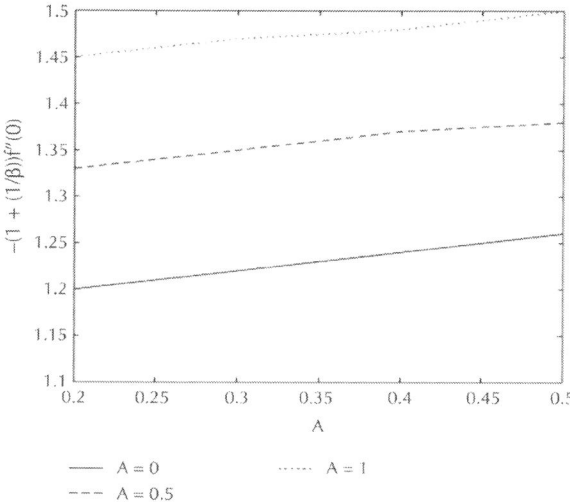

(b) Skin friction versus unsteadiness parameter at different values of the permeability parameter A at $M = 1, \beta = 2, Pr = 1, Sc = 1,$ and $R = 0.5$

**Figure 6:** Variation of (a) skin friction coefficient with unsteadiness parameter for different $\beta$ values and (b) with different $\Lambda$ values.

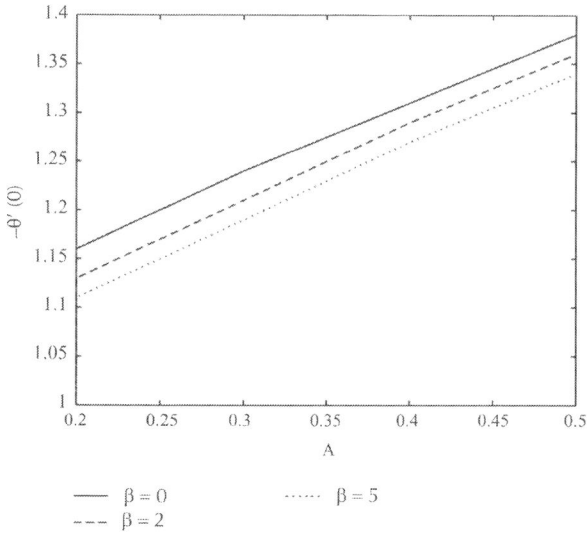

(a) Heat transfer coefficient versus unsteadiness parameter at different values of the Casson parameter b at $M = 1, A = 0.5, Or = 1, Sc = 1$ and $R = 0.5$

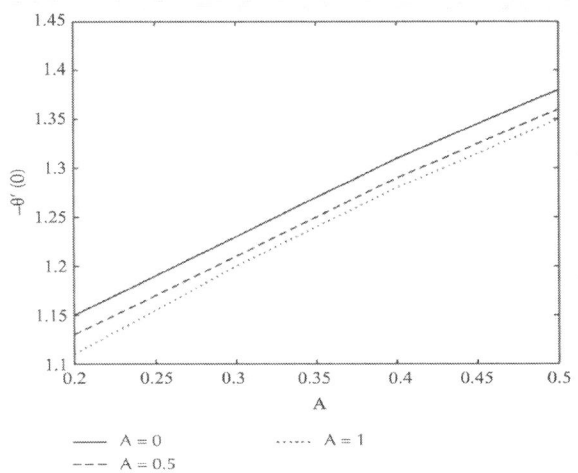

(b) Heat transfer coefficient versus unsteadliness parameter at different values of the permeability paramerter A at M = 1, b = 0.5, Pr = 1, Sc = 1, amd R = 0.5

**Figure 7:** Variation of (a) heat transfer coefficient with unsteadiness parameter for different $\beta$ values and (b) with different $\Lambda$ values.

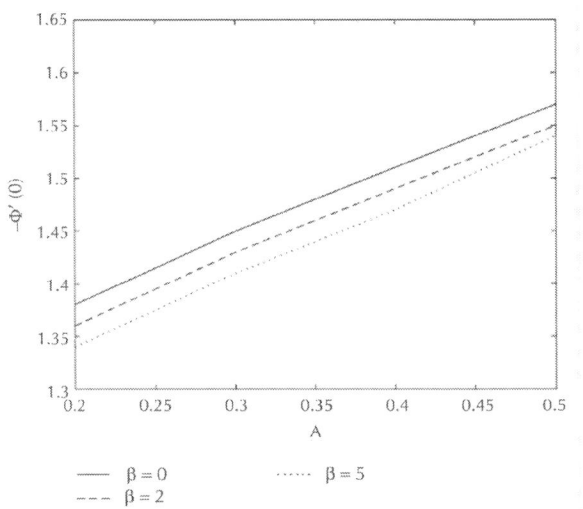

(a)Mass transfer coefficient versus unsteadiness parameter at differernt values of the Cassson parameter $\beta$ at M=1,A=0.5, Pr = 1, Sc =1, and R = 0.5

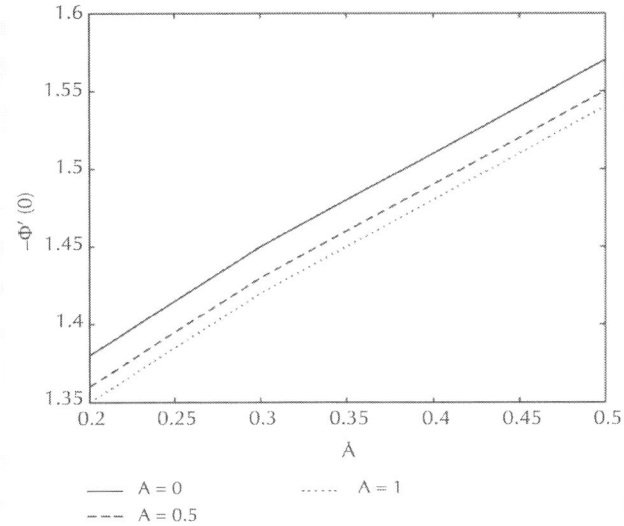

(b) Heat transfer coefficient versus unsteadiness parameter at differernt values of the permeability parameter A at M =1, β = 2, Pr = 1, Sc =1 Sc and R = 0.5

**Figure 8:** Variation of (a) mass transfer coefficient with unsteadiness parameter for different $\beta$ values and (b) with different $\Lambda$ values.

# CONCLUSIONS

The study presented in this analysis of diffusion of chemically reactive species in Casson fluid flow over an unsteady stretching surface in a porous medium in the presence of a magnetic field provides numerical solutions for the boundary layer flow and heat and mass transfer. The coupled nonlinear governing equations were solved using the Runge-Kutta-Fehlberg integration scheme. Increasing the unsteadiness parameter decreases velocity profiles. Increasing the Casson parameter decreases the velocity profiles. The presence of heat transfer does not seem to significantly change the results obtained by Mukhopadhyay and Vajravelu [1]; the main difference is noted in the presence of magnetic and permeability parameters which they did not consider. In this study it is generally noted that increasing the magnetic parameter and permeability parameter decreases the velocity profiles but increases the skin friction; it then decreases the coefficient of heat transfer and the concentration profiles.

# REFERENCES

1. S. Mukhopadhyay and K. Vajravelu, "Diffusion of chemically reactive species in Casson fluid flow over an unsteady permeable stretching surface," Journal of Hydrodynamics, vol. 25, no. 4, pp. 591–598, 2013.
2. S. Mukhopadhyay, P. R. de, K. Bhattacharyya, and G. C. Layek, "Casson fluid flow over an unsteady stretching surface," Ain Shams Engineering Journal, vol. 4, no. 4, pp. 933–938, 2013.
3. M. Abd El-Aziz, "Mixed convection flow of a micropolar fluid from an unsteady stretching surface with viscous dissipation," Journal of the Egyptian Mathematical Society, vol. 21, no. 3, pp. 385–394, 2013.
4. L. J. Grubka and K. M. Bobba, "Heat transfer characteristics of a continuous stretching surface with variable temperature," Journal of Heat Transfer, vol. 107, no. 1, pp. 248–250, 1985.
5. S. Sharidan, T. Mahmood, and I. Pop, "Similarity solutions for the unsteady boundary layer flow and heat transfer due to a stretching sheet," International journal of Applied Mechanical Engineering, vol. 11, pp. 647–654, 2006.
6. S. Nadeem, R. Ul Haq, and C. Lee, "MHD flow of a Casson fluid over an exponentially shrinking sheet,"Scientia Iranica, vol. 19, no. 6, pp. 1550–1553, 2012.
7. S. Nadeem, R. U. Haq, and Z. H. Khan, "Numerical study of MHD boundary layer flow of a Maxwell fluid past a stretching sheet in the presence of nanoparticles," Journal of the Taiwan Institute of Chemical Engineers, vol. 45, no. 1, pp. 121–126, 2014.
8. K. Ahmed and R. Nazar, "Magnetohydrodynamic three dimensional flow and heat transfer over a stretching surface

in a viscoelastic fluid," Journal of Science and Technology, vol. 3, no. 1, pp. 33–46, 2011.

9. K. Rohlf and G. Tenti, "The role of the Womersley number in pulsatile blood flow a theoretical study of the Casson model," Journal of Biomechanics, vol. 34, no. 1, pp. 141–148, 2001.

10. D. S. Sankar and U. Lee, "Two-fluid non-linear model for flow in catheterized blood vessels,"International Journal of Non-Linear Mechanics, vol. 43, no. 7, pp. 622–631, 2008.

11. D. S. Sankar and U. Lee, "Two-fluid Casson model for pulsatile blood flow through stenosed arteries: a theoretical model," Communications in Nonlinear Science and Numerical Simulation, vol. 15, no. 8, pp. 2086–2097, 2010.

12. S. Shaw, R. S. R. Gorla, P. V. S. N. Murthy, and C.-O. Ng, "Pulsatile casson fluid flow through a stenosed bifurcated artery," International Journal of Fluid Mechanics Research, vol. 36, no. 1, pp. 43–63, 2009.

13. A. V. Mernone, J. N. Mazumdar, and S. K. Lucas, "A ematical study of peristaltic transport of a Casson fluid," ematical and Computer Modelling, vol. 35, no. 7-8, pp. 895–912, 2002.

14. B. Das and R. L. Batra, "Secondary flow of a Casson fluid in a slightly curved tube," International Journal of Non-Linear Mechanics, vol. 28, no. 5, pp. 567–577, 1993. ·

15. C.-O. Ng, "Combined pressure-driven and electroosmotic flow of Casson fluid through a slit microchannel," Journal of Non-Newtonian Fluid Mechanics, vol. 198, pp. 1–9, 2013.

16. S. Nadeem, R. Ul Haq, N. S. Akbar, and Z. H. Khan, "MHD three-dimensional Casson fluid flow past a porous linearly stretching sheet," Alexandria Engineering Journal, vol. 52, no. 4, pp. 577–582, 2013.

17. R. K. Dash, K. N. Mehta, and G. Jayaraman, "Casson fluid flow in a pipe filled with a homogeneous porous medium," International Journal of Engineering Science, vol. 34, no. 10, pp. 1145–1156, 1996.

18. D. Tripathi, "Study of transient peristaltic heat flow through a finite porous channel," ematical and Computer Modelling, vol. 57, no. 5-6, pp. 1270–1283, 2013. ·
19. S. Pramanik, "Casson fluid flow and heat transfer past an exponentially porous stretching surface in presence of thermal radiation," Ain Shams Engineering Journal, vol. 5, no. 1, pp. 205–212, 2014.
20. P. V. Ramachandra, R. A. Subba, and B. O. Anwa, "Flow and heat transfer of casson fluid from a horizontal cylinder with partial slip in non-darcy porous medium," Applied and computational ematics, vol. 2, article 2, 2013.
21. P. K. Kameswaran, S. Shaw, P. Sibanda, and P. V. S. N. Murthy, "Homogeneous-heterogeneous reactions in a nanofluid flow due to a porous stretching sheet," International Journal of Heat and Mass Transfer, vol. 57, no. 2, pp. 465–472, 2013.
22. S. Shaw, P. K. Kameswaran, and P. Sibanda, "Homogeneous-heterogeneous reactions in micropolar fluid flow from a permeable stretching or shrinking sheet in a porous medium," Boundary Value Problems, vol. 2013, article 77, 2013.
23. A. J. Chamkha, A. M. Aly, and M. A. Mansour, "Similarity solution for unsteady heat and mass transfer from a stretching surface embedded in a porous medium with suction/injection and chemical reaction effects," Chemical Engineering Communications, vol. 197, no. 6, pp. 846–858, 2010.
24. G. Makanda, O. D. Makinde, and P. Sibanda, "Natural convection of viscoelastic fluid from a cone embedded in a porous medium with viscous dissipation," ematical Problems in Engineering, vol. 2013, Article ID 934712, 11 pages, 2013.

# Encapsulation of Menthol in Beeswax by a Supercritical Fluid Technique

Linjing Zhu[1], Hongqiao Lan[2], Bingjing He[1], Wei Hong[1], and Jun Li[1]

[1]Department of Chemical and Biochemical Engineering, College of Chemistry and Chemical Engineering, Xiamen University, Xiamen 361005, China

[2]Technology Center, China Tobacco Fujian Industry Corporation, Xiamen 361022, China

## ABSTRACT

Encapsulation of menthol in beeswax was prepared by a modified particles from gas-saturated solutions (PGSS) process with controlling the gas-saturated solution flow rate. Menthol/beeswax particles with size in the range of 2–50 μm were produced. The effects of the process conditions, namely, the pre-expansion pressure, pre-expansion temperature, gas-saturated solution flow rate, and menthol composition, on the particle size, particle size distribution, and menthol encapsulation rate were investigated. Results indicated that in the range of studied conditions, increase of the pressure, decrease of the gas-saturated solution flow rate, and decrease of the menthol mass fraction can decrease the particle size and narrow particle size distribution of the produced menthol/beeswax microparticles. An $N_2$-blowing method was proposed to measure the menthol release from the menthol/beeswax microparticles. Results showed that the microparticles have obvious protection of menthol from its volatilization loss.

# INTRODUCTION

Natural menthol exists in peppermint and other mint oils, having minty odor. It is widely used as tobacco flavor to manufacture menthol cigarettes, inhibit stimulation from tobacco, and cover miscellaneous gases. However, the melting temperature of menthol is 42~43°C at atmospheric pressure, which makes it easy to sublimate. The high volatility not only means its short storage time and its pollution to the work environment and product line in the menthol cigarette production process but also suggests the flavor's rapid loss with smoking. Obviously, the control and detection of menthol transfer is an important issue during the production, storage, and smoking of menthol cigarette. The encapsulation of menthol will not only provide coolness and refreshing taste from menthol but will also ensure controlled release of menthol; this may also improve the effectiveness, broaden the time range of menthol flavor, and ensure optimal dosage [1]. Hee [2] studied the transfer of menthol from tobacco to filter and the menthol percentages of different menthol cigarettes, indicating that adding menthol to the plate of paper instead of tobacco is the most effective method. Peng [3] studied the menthol composite particles by using phase separation-condensation method, and results showed that the composite particles with an encapsulation efficiency of menthol up to 58.7% could not only enhance the retention of menthol in the cigarette but also improve the smoke quality.

In this study, the used wall material is beeswax, which has special fragrance of honey and flour. It can melt at 62~67°C at atmospheric pressure. Furthermore, beeswax completely turns to liquid in the condition of 15MPa and 55.2°C as tested by using laboratory equipment [4].

Conventional methods for preparing composite particles such as spray drying, spray cooling, spray chilling, air-suspension coating, extrusion, centrifugal extrusion, freeze drying, coacervation, rotational suspension separation, cocrystallization, and interfacial polymerization often involve the use of organic solvents, thermal and chemical degradation of compounds, high-energy requirement, multistage processes, environmental pollution, low encapsulation efficiency, and broad particle size distribution. These limitations are the main reasons for the increasing interest in effective and clean encapsulation methods [5]. Particles from gas-saturated solutions (PGSS) process proposed by Weidner et al. [6] employed supercritical fluid technology to prepare microparticles. Currently, various studies have focused on the composite particles prepared by using this technique. For example, Rodrigues et al. [7] prepared microcomposites theophylline/hydrogenated palm oil from a conventional PGSS process and found that limited theophylline was encapsulated in hydrogenated palm oil. Wang et al. [8] investigated lipid/ibuprofen system and controlled release is obviously achieved by the process. Vezzù et al. [9] with a PGSS studied the encapsulation of magnetite nanoparticles (MNPs) into triestearin and phosphatidylcholine mixtures to increase their biocompatibility for future applications in the fields of biomedical diagnostics and therapeutic medications. Recently, Pemsel et al. [10] studied the encapsulation of the biopesticide Cydia pomonella granulovirus by PGSS, and the wall materials were an organic matrix together with admixed UV protectants and phagostimulants. Due to the low operation temperature in the process, in particular, the low temperature at the tip of the nozzle exit, PGSS favors food industry, for example, for preventing flavors from volatilizing. In addition, the technique does not need to use water or organic solvents.

Due to the high volatility of menthol, this research will use a modified PGSS process which can control the flow rate of the gas-saturated solution to prepare the menthol/beeswax particles aimed at preventing menthol from volatilization loss. The influences of the pre-expansion pressure, pre-expansion temperature, gas-saturated solution flow rate and menthol composition, on the particle size, particle size distribution, and menthol encapsulation efficiency will be investigated. Also, a $N_2$-blowing method will be used to test the menthol release of the microparticles.

# EXPERIMENTAL

## Materials

Menthol (CP, purity ≧99%) and beeswax (mixture) were purchased from Xiamen Glass Instrument Co., Ltd. Greenery Reagent. $CO_2$ was provided by Tongan Air Separation & Special Gas Factory, Xiamen, China. Anhydrous ethanol (analytical grade) and Ether (analytical grade) were provided from the first Chemical Plant of Shanghai Zhenxing. Acetophenone (analytical grade) was provided from Xiamen Glass Instrument Co., Ltd. Greenery Reagent.

## Apparatus and Procedure

The modified PGSS process used in this experiment is shown in Figure 1. As it shows, the major difference of the process from conventional PGSS process is the addition of the two high-pressure pumps (E1 and E2) for delivering gas-saturated solution. The advantage of the delivery of the gas-saturated solution instead of charge of melted solute solution (unsaturated solution before mixing with high pressure $CO_2$) is evident: (1) the process can be operated at relatively low temperature due to the depression of the melting temperature of solid solute in high pressure of $CO_2$; (2) the block problem in the nozzle in conventional PGSS process can be effectively reduced with controlling the flow rate of the gas-saturated solution and the corresponding coaxial nozzle designed (a pipe with O.D of 3 mm in a pipe with I.D. of 4 mm; the gas-saturated solution passes through the inner pipe and $CO_2$ goes through the gap between the inner and outer pipes); (3) the solute can be mixed with $CO_2$ efficiently by using E2 to form gas-saturated solution.

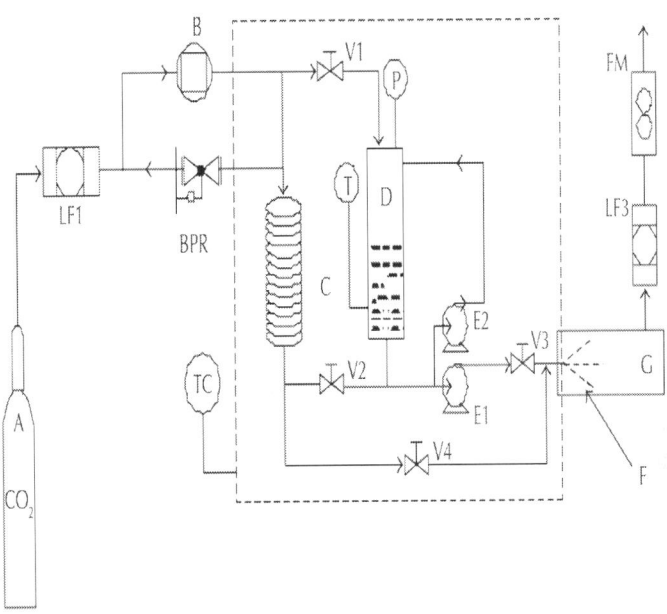

**Figure 1:** Schematic diagram of the modified PGSS process. A: $CO_2$ cylinder; B: compressor; C: preheater; D: mixer; E#: high-pressure pump; F: double-pass nozzle; G: precipitant; P: pressure gauge; T: thermometer; V#: valve; FM: flow meter; TC: temperature controller; LF#: filter; BPR: back pressure valve.

The procedure of the process is described as follows.
- $CO_2$ is compressed into the gas damper to the desired pre-expansion pressure and enters the thermostatic system (the

operating temperature is controlled by the air bath with a precision of ±0.1°C). The compressed $CO_2$ is divided into two parts: one passes through the preheater C and valve 4 (which was closed before preparing particles), then enters directly into the external tube of the nozzle system; another passes through valve 1 into the high-pressure mixing vessel and mixes with solid materials added before the experiment to form gas-saturated solution.

- The gas-saturated solution is pumped by E2 into the vessel for cycling and mixing with a relatively large flow rate (such as more than 5~10 mL/min) in the case of fast mixing.
- After mixing, E1 (E1 and E2 can be combined together if only one pump is available) works after opening valve 3 with a flow rate of 0.1~1 mL/min (other pumps can be chosen for larger or smaller flow rates) to charge the gas-saturated solution into the inner tube of the nozzle system. This gas-saturated solution is atomized by the $CO_2$ from the external tube through a disc F with a laser-drilled orifice of 80 μm into the collector G to form fine particles. The overall flow rate of $CO_2$ exhausted is measured by a gas flow meter after filtration. All the particles in the collector were taken out after the experiment and stored in a sealed bottle; the encapsulation efficiency of the sample (arbitrarily sampled from the collected particles) was measured right after the collection for a correct indication of the content of menthol on the surface of the particles.

With the use of menthol as the core material and beeswax as the wall material, the above process was implemented. According to the melting point tested for beeswax (the melting point of the mixture is lower than that of beeswax or menthol), the operating temperature is fixed at 60°C. The orifice diameter of 80 μm was selected for convenient control of the process (larger-size nozzle can cause difficulty to maintain a high pressure). The influences of the operating pressure $P_0$, flow rate L and menthol mass fraction $C_0$ on the formation of the menthol/beeswax particles were investigated; the experiment was arranged as Table 1 shows. Note that the last column provides the measured menthol mass fraction

(C) in the produced menthol/beeswax particles, which includes the menthol on the particles surface and in the particles. Table 1 shows that C is very close to $C_0$ with a minimum deviation of the two mass fractions of 0.0% and a maximum deviation of 2.4%.

**Table 1:** Experiment conditions

| Run No. | $P_0$, MPa | L, mL/min | $C_0$ | C |
|---|---|---|---|---|
| 1 | 6 | 0.21 | 10% | 10.5% |
| 2 | 10 | 0.21 | 10% | 9.8% |
| 3 | 15 | 0.21 | 10% | 10.6% |
| 4 | 20 | 0.21 | 10% | 11.7% |
| 5 | 15 | 0.21 | 20% | 20.0% |
| 6 | 15 | 0.21 | 30% | 32.1% |
| 7 | 15 | 0.21 | 40% | 38.4% |
| 8 | 15 | 0.11 | 10% | 12.4% |
| 9 | 15 | 0.40 | 10% | 11.2% |
| 10 | 15 | 0.81 | 10% | 12.1% |

$P_0$: opterating pressure; $T_0$: operating temperature = 60°C; D: nozzle size = 80 μm; L: the flow rate of solution (calibrated with pure water at high pressure); $C_0$: mass fraction of menthol in the menthol/beeswax mixture; C: measured menthol mass fraction in the produced menthol/beeswax particles.

## Analysis Methods

The morphology and size of the menthol/beeswax particles were analyzed by XL30 environment scanning electron microscope (SEM). The particle size and size distribution were analyzed by a laser particle size analyzer (LS908, OMEC, China) with static method.

The menthol encapsulation efficiency in the menthol/beeswax particles is calculated by

$$\text{encapsulation efficiency (\%)} = \frac{\text{total menthol (g)} - \text{menthol on the surface (g)}}{\text{total menthol (g)}} \times 100, \tag{1}$$

where the menthol content on the surface of the particles and the total menthol content were measured by gas chromatography (GC) (FID as the detector; SE-54 column: 30 m × 0.32 mm × 0.25 µm; nitrogen as the carrier gas with a flow rate of 1 mL/min; column temperature: 70°C for 1 min, then 5°C/min, and finally 140°C for 1 min; injection port temperature 250°; detector temperature 250°C; split ratio 50:1; injection volume 0.4 µL). Ethanol was used to dissolve the menthol on the surface of the produced particles and ether was used to dissolve the total menthol of the particles.

During the storage of the menthol/beeswax particles, the menthol in the particles will escape via volatilization because of the influence of the external environment. So, the release of the core material is an important indication for evaluating the composite particles, which is calculated by

$$\text{menthol release (\%)} = \frac{\text{menthol content (g) at time } t}{\text{initial menthol content (g)}} \times 100, \tag{2}$$

where the initial content of menthol measured by GC is indicated in Table 1. The menthol content at time t is measured by a $N_2$-blowing method: the menthol/beeswax particles sealed in a U-tube, $N_2$ with a pressure of 0.15 MPa passes through the sample in the glass tube; the U-tube is weighed after an interval time to obtain the released amount of menthol in the particles.

# RESULTS AND DISCUSSION

## The Influence of Pre-Expansion Pressure

The effects of four pre-expansion pressures (6, 10, 15, and 20 MPa) were investigated on the formed menthol/beeswax particles when other conditions were fixed ($C_0$=10%; L=0.21 mL/min; $T_0$=60°C; D=80 µm). Figure 2 gives the average particle size, particle size distribution (PSD), and the encapsulation efficiency of the produced particles. Because the produced particles have very similar morphology, Figure 2(d) provides only a typical SEM image for the menthol/beeswax particles produced at 10 MPa (because of the low melting temperature of the particles, SEM with small scale cannot be performed).

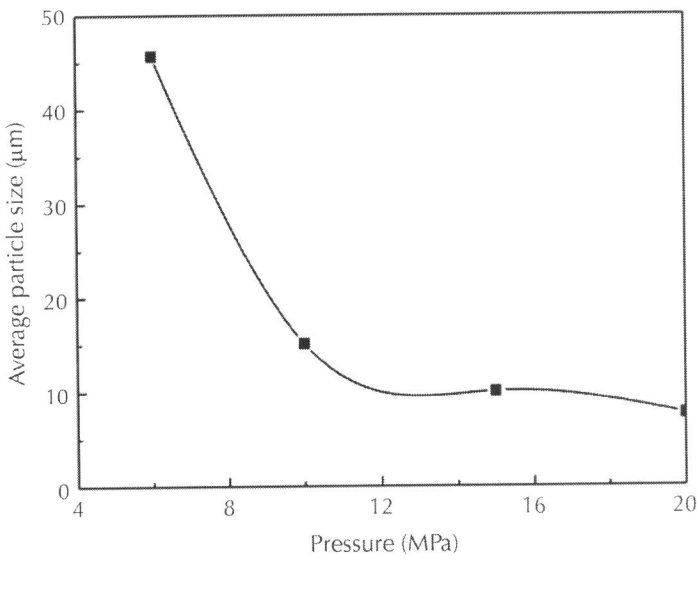

(a)

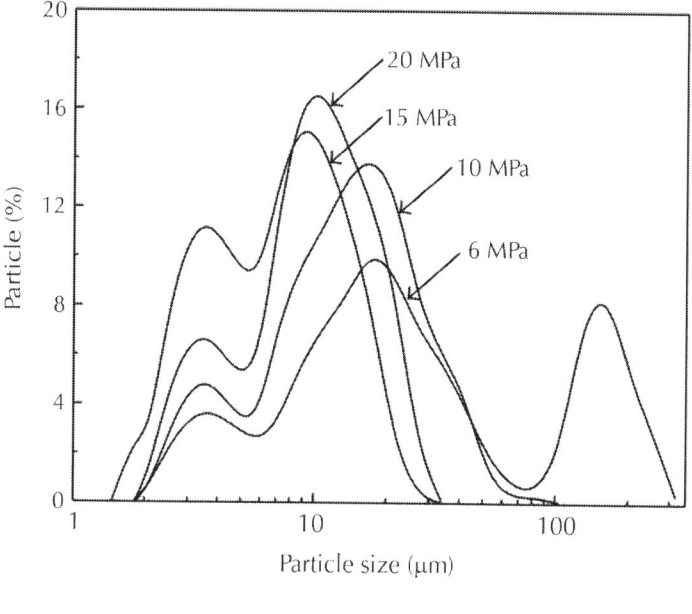

(b)

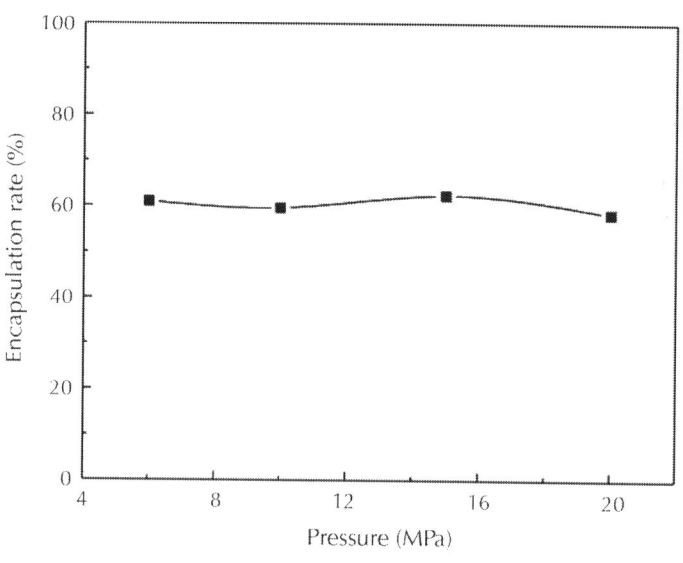

(c)

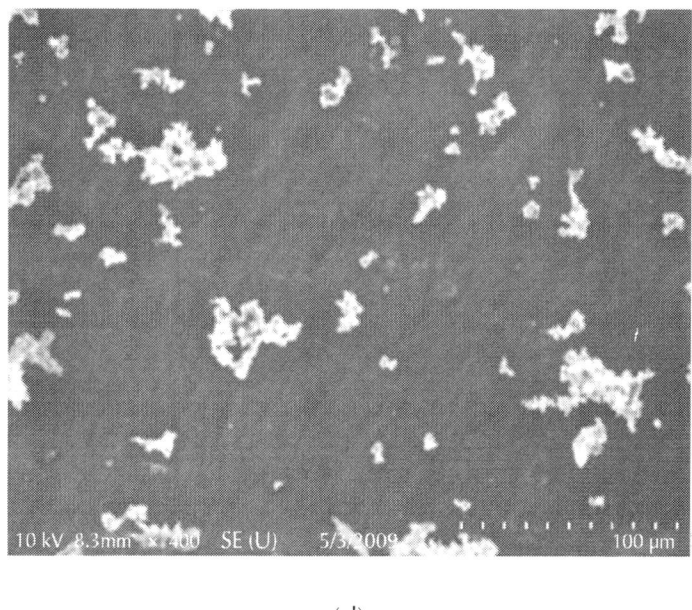

(d)

**Figure 2:** Influence of the pre-expansion pressure on (a) average particle size, (b) PSD, (c) encapsulation efficiency, and (d) SEM image of the menthol/beeswax particles produced at 10 MPa.

As shows in Figure 2, as the pressure increases, the average particle size decreases, and particle size distribution becomes narrow. The menthol/beeswax particles prepared at 6 MPa were the largest with an average size of 45 μm due to an additional peak appearing at 180 μm. This mainly attributes to particle aggregation, because the temperature drop at the nozzle exit was not enough when low pressure of $CO_2$ is applied, leading to relatively slow solidification of the menthol/beeswax mixture. The menthol/beeswax particles prepared at other pressures have two peaks (bimodal distribution) obviously from different particle formation mechanisms [11]. Lower pressure tends to diminish the first peak corresponding particles considered to be precipitated from $CO_2$ phase, which is in agreement with the previous modeling conclusions [11]. Figure 2(c) shows that the encapsulation efficiency of all the menthol/beeswax particles is about 60%, meaning that the pre-expansion pressure has no effect on the encapsulation efficiency. Since the encapsulation efficiency is approximately 60% for all particles with different sizes, this result may also suggest that menthol could be dispersed into beeswax in a relative homogeneous way. In addition, the agreement of the measured menthol content via GC, including that on the particles surface and in the particles, in the produced menthol/beeswax particles arbitrarily sampled and the mass fraction of menthol in the initial menthol/beeswax mixture (see the last two columns in Table 1) also supports the relative homogeneity of the produced particles.

## The Influence of the Flow Rate of Solution

As mentioned, the obvious advantage of the modified PGSS process is the control of the gas-saturated solution. Consequently, different solution flow rates (0.11, 0.21, 0.40, and 0.81 mL/min) were investigated, while other conditions were fixed ($C_0=10\%$; $P=15$ MPa; $T_0=60°C$; $D=80$ μm). Figure 3 provides the average particle size, PSD, and the encapsulation efficiency of the produced menthol/beeswax particles.

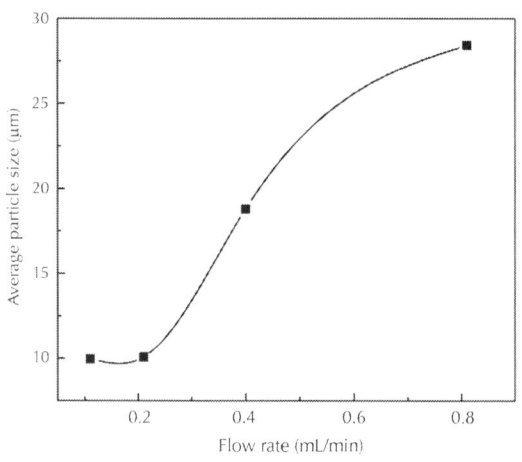

(a)

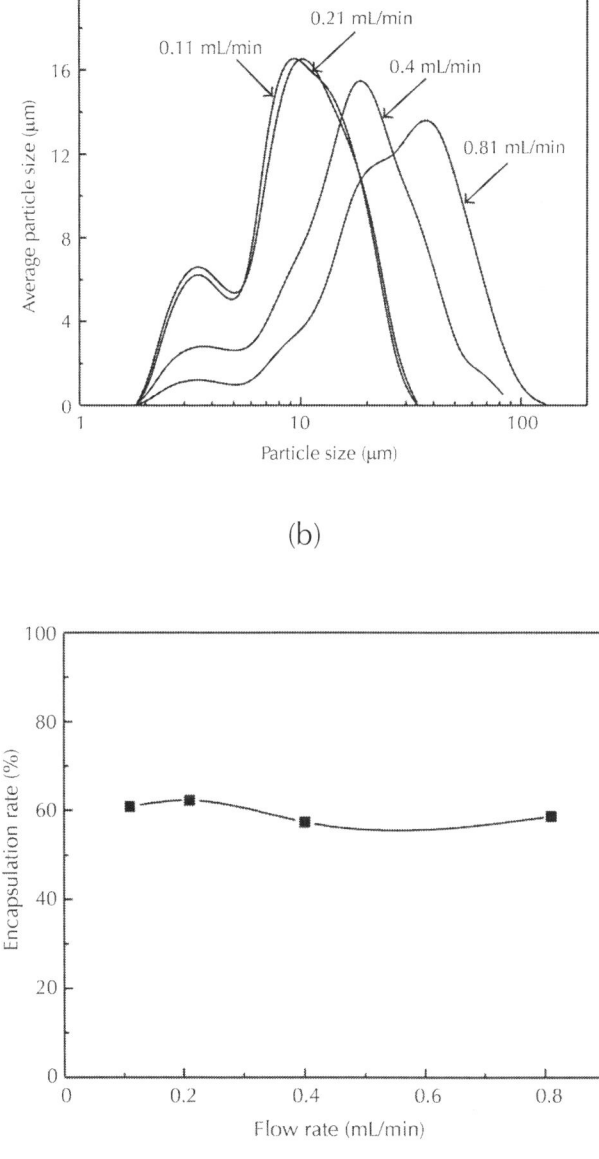

**Figure 3:** Influence of the flow rate on (a) average particle size, (b) PSD, and (c) encapsulation efficiency.

As shown in Figure 3, the flow rate has an obvious effect on the produced particles: as the flow rate increases, the average particle size increases, and PSD becomes broadened. The menthol/beeswax particles produced at lower flow rate such as 0.11 and 0.21 mL/min have very close particle size and similar PSD (bimodal distribution). Yet, the menthol/beeswax particles produced at larger flow rate such as 0.40 and 0.81 mL/min tend to diminish the first peak, indicating that atomization mechanism is in dominance at lower solution rate (but this factor was not included in the modeling [11] because the model is only for a semibatch PGSS process). Similarly, Figure 3(c) shows that the encapsulation efficiency of all the menthol/beeswax particles is about 60%, indicating that the flow rate has no obvious effect on the encapsulation efficiency.

## The Influence of the Initial Menthol Content

As the menthol content will affect the encapsulation efficiency, its effect on the formed menthol/beeswax particles was investigated. Four different initial menthol mass fractions, namely, 10%, 20%, 30%, and 40%, were implemented with fixing the pre-expansion pressure of 15 MPa and the flow rate of 0.11 mL/min. Figure 4 provides the average particle size, PSD, and the encapsulation efficiency of the produced menthol/beeswax particles.

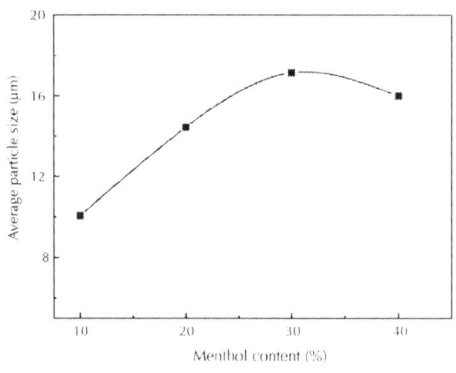

(a)

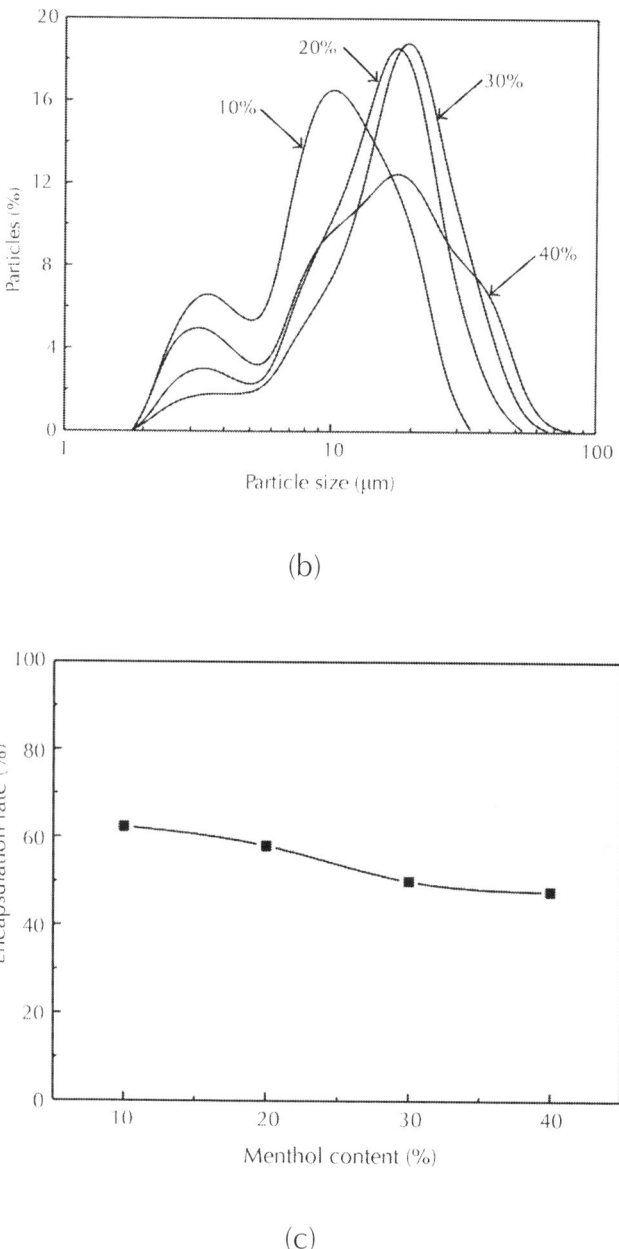

**Figure 4:** Influence of the initial menthol content on (a) average particle size, (b) PSD, and (c) encapsulation efficiency.

As shown in Figure 4, as the menthol content increases, the average particle size slightly increases, and the effect disappears when the mass fraction arrives at about 30%. Yet, the effect of the menthol content on the encapsulation efficiency of the composite particles is obvious: higher menthol concentration causes to lower encapsulation efficiency. This is reasonable because higher menthol mass fraction in the initial menthol/beeswax mixture will cause more menthol to stay on the surface of the produced particles.

## Release Curve

For the $N_2$-blowing method, only the menthol/beeswax particles produced at $P_0=10$ MPa; $C_0=10\%$ L=0.21 mL/min; $T_0=60°C$; D=80 μm (see Figure 2(d) for the particles) were tested. A U-tube loaded with 0.1 g sample and immersed in a water bath at 30°C was connected to a $N_2$ cylinder with a reducing valve; the $N_2$ (0.15 MPa) passed through the sample to obtain the retention of menthol in the menthol/beeswax particles at different interval times. Figure 5 gives the menthol release from the particles, and it also provides the menthol release from a physical mixture of menthol/beeswax with $C_0=10\%$ for comparison. The figure shows that after the purge of $N_2$ for about 80 h, the residue of menthol in the menthol/beeswax particles steadily stops at 54%. This residue is consistent with the measured encapsulation rate (59.6%) indicated in Figure 2(c) for the same sample, indicating that the $N_2$-blowing experiment is also available for measuring the encapsulation efficiency. From the release curve, it clearly shows that the menthol on the surface of the menthol/beeswax particles will gradually run away and cannot be protected for long time. On the other hand, the menthol in the physical mixture loses quickly (90% of menthol in the mixture releases in 20 h); the slow release of the left menthol after 20 h (namely, only about 5% of menthol releases during 20 to 40 h) may be attributed to the adsorption of the menthol on the surface of beeswax particles.

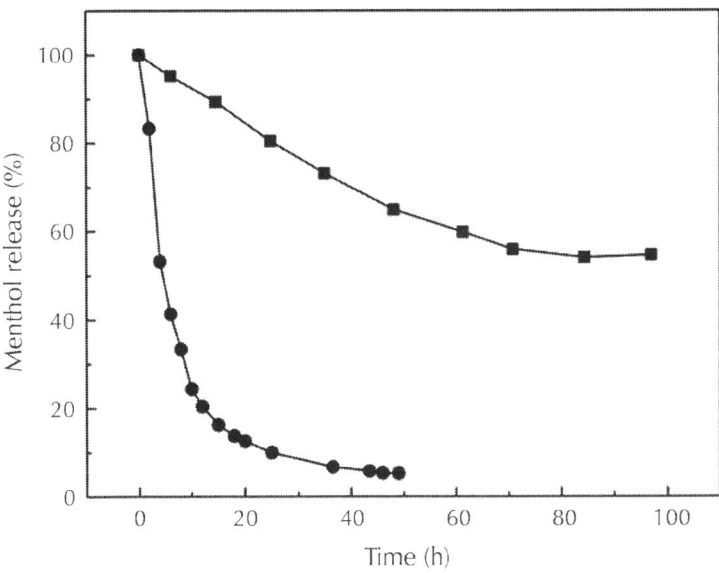

**Figure 5:** Menthol release from the microparticles (■) and physical mixture (●) of menthol/beeswax.

## CONCLUSIONS

A modified PGSS process with control of gas-saturated solution flow rate was used to produce menthol/beeswax microparticles in order to protect menthol from volatilizing. The effect of the process parameters, namely, the pre-expansion pressure, the gas-saturated solution flow rate, and the menthol mass fraction in menthol/beeswax mixture, on the particle size, particle size distribution, and menthol encapsulation efficiency was investigated. Furthermore, a $N_2$-blowing method was proposed to measure the menthol release from the produced menthol/beeswax microparticles. From these studies, the following conclusions can be drawn.
- The modified PGSS process can be employed to prepare menthol/beeswax microparticles with size in the range of 2~50 µm. The menthol content measured by GC in the

- produced particles is consistent with the initial menthol mass fraction in the menthol/beeswax mixture, indicating uniform composite particles were obtained by the process.
- Both the pre-expansion pressure and gas-saturated solution flow rate have evident influence on the size of composite particles: as the pre-expansion pressure increases, the gas-saturated solution flow rate decreases, the average size of menthol/beeswax particles decreases and the particle size distribution becomes narrow. Both the pre-expansion pressure and gas-saturated solution flow rate have no obvious influence on the encapsulation efficiency of menthol.
- As the menthol mass fraction increases, the average particle size slightly increases, and the encapsulation efficiency of menthol decreases.
- The $N_2$-flush measurement for the release of menthol from the menthol/beeswax particles indicates that the microparticles have obvious protection of menthol from its volatilization loss although the menthol on the menthol/beeswax particles surface will volatilize gradually. The $N_2$-flush technique is also available for measuring the encapsulation efficiency.

# ACKNOWLEDGMENTS

For financial support, the authors are grateful to SRF for ROCS, SEM, NCET of Fujian Province, NSFC, China (Project no. 20876127), and Tobacco S&T Major Project of China National Tobacco Corporation (Project no. 11020092021).

# REFERENCES

1. S. Gouin, "Microencapsulation: industrial appraisal of existing technologies and trends," Trends in Food Science and Technology, vol. 15, no. 7-8, pp. 330–347, 2004.

2. J. J. Hee, "A study on menthol migration patternsin different mentholated cigarettes,"Journal of the Korea Society of Tobacco Science, vol. 23, pp. 77–81, 2001.
3. R. H. Peng, "Menthol composite particles prepared by phase separation-coacervation method experiment," Tobacco Science, vol. 8, pp. 27–28, 2003.
4. J. Li, M. Rodrigues, A. Paiva, H. A. Matos, and E. G. DeAzevedo, "Binary solid-liquid-gas equilibrium of the tripalmitin/$CO_2$ and ubiquinone/$CO_2$ systems," Fluid Phase Equilibria, vol. 241, no. 1-2, pp. 196–204, 2006.
5. I. Garay, A. Pocheville, and L. Madariaga, "Polymeric microparticles prepared by supercritical antisolvent precipitation," Powder Technology, vol. 197, no. 3, pp. 211–217, 2010.
6. E. Weidner, R. Steiner, Z. Knez, and Z. Novak, "Powder generation form polyethyleneglycols with compressible fluids," High Pressure Chemical Engineering, vol. 12, pp. 223–228, 1996.
7. M. Rodrigues, N. Peiriço, H. Matos, E. Gomes De Azevedo, M. R. Lobato, and A. J. Almeida, "Microcomposites theophylline/hydrogenated palm oil from a PGSS process for controlled drug delivery systems," Journal of Supercritical Fluids, vol. 29, no. 1-2, pp. 175–184, 2004.
8. X. Wang, Y. N. Guo, H. Chen, et al., "Composite microparticles of ibuprofen/lipid generated by supercritical fluids from their melts," Frontiers of Chemical Engineering in China, vol. 2, no. 4, pp. 361–367, 2008.
9. K. Vezzù, C. Campolmi, and A. Bertucco, "Production of lipid microparticles magnetically active by a supercritical fluid-based process," International Journal of Chemical Engineering, vol. 2009, Article ID 781247, 9 pages, 2009. ·
10. M. Pemsel, S. Schwab, A. Scheurer, D. Freitag, R. Schatz, and E. Schlücker, "Advanced PGSS process for the encapsulation of the biopesticide Cydia pomonella granulovirus,"Journal of Supercritical Fluids, vol. 53, no. 1–3, pp. 174–178, 2010.

11. J. Li, M. Rodrigues, A. Paiva, H. A. Matos, and E. G. De Azevedo, "Modeling of the PGSS process by crystallization and atomization," AIChE Journal, vol. 51, no. 8, pp. 2343–2357, 2005.

# Chapter 4

# Direct Solid-State Fermentation of Soybean Processing Residues for the Production of Fungal Chitosan by Mucor Rouxii

Andro Mondala, Ramea Al-Mubarak, James Atkinson, Shaun Shields, Brian Young, Yurguen Dos Santos Senger, and Jan Pekarovic

Department of Chemical and Paper Engineering, Western Michigan University, Kalamazoo, USA

## ABSTRACT

The feasibility of utilizing soybean-processing residues such as soybean meal and hulls as substrates for chitosan production by the fungus Mucor rouxii ATCC 24905 via solid-state fermentation (SSF) was investigated. The effects of the type of soybean-based substrate, length of cultivation period, substrate moisture content, substrate pH, incubation temperature and extraction conditions on chitosan yield were determined. The results showed that a maximum fungal chitosan yield of up to 3.44% by dry substrate weight (34.4 g/kg) could be achieved using a pure soybean meal substrate with an initial moisture content of 50% (w/w) and pH of 5 - 6 incubated for six days at 25°C. A more severe heat treatment (autoclaving vs. refluxing) resulted in higher chitosan extraction yields regardless of the strength of extraction reagents. Fourier transform infrared (FTIR) analysis of the fungal chitosan revealed its degree of deacetylation (DDA) to be between 55% and 60%.

# INTRODUCTION

Chitosan is a natural biopolymer that has gained increased attention in recent years due to its interesting and beneficial biological and molecular characteristics, which enable its application in a wide variety of industrial and biotechnological fields. These fields range from agriculture, environmental remediation, pulp and paper coatings and barrier films, food preservation, and cosmetics to more advanced technology emerging fields such as functional materials in flexible printed organic electronic devices, protein separations, drug delivery, and tissue engineering [1] . Chitosan is a heterogeneous polysaccharide containing $\beta$-(1-4)-linked 2-amino-2-deoxy-D- glucose (glucosamine) and N-acetyl-2-amino-2-deoxy-D-glucose (N-acetyl glucosamine) units, with the former consisting of more than 60 % of the copolymer [2] . It is obtained through deacetylation of chitin, which contains mostly N-acetyl glucosamine units. Its unique and beneficial properties enabling its numerous applications include its cationic nature, chelation and ion-binding ability, protein immobilization, film- and gel-forming characteristics, chemical reactivity, amenability to modification, biocompatibility, and antimicrobial activity [3] . These properties are a consequence of its heterogeneous molecular structure and the presence of highly reactive amine groups at the glucosamine residues [4] . Fields and applications that can be served by utilizing chitosan and exploiting its properties include biomedical engineering, food and agriculture, environmental remediation, and polymers, films, fibers and coatings [1] - [3].

Chitosan is traditionally produced by subjecting crustacean shells to severe alkali, acid, and/or enzymatic treatments to isolate chitin, by removing impurities such as proteins, minerals and pigments, and by deacetylating of chitin to chitosan—that is, the removal of the acetyl groups in the N-acetyl glucosamine units comprising the majority of chitin to increase the proportion of glucosamine units in the polymer. Harsh reaction conditions are required due to the complex organization of chitin with other components in the crustacean exoskeleton [5]. Thus, the conventional process results in high labor and processing costs. Additionally, these raw materials are seasonally and geographically limited and have highly variable characteristics, which can lead to inconsistent product yield and molecular characteristics. This is undesirable for high purity applications such as in the pharmaceutical and biomedical fields [3] [6]. On the other hand, chitosan is also found as a major fungal cell wall component particularly of fungal strains belonging to the order Mucorales of the Zygomycetes class [6] [7]. These fungal species produce chitosan directly unlike crustacean-derived chitosan, which has to be produced via isolation and chemical conversion of chitin from the exoskeletons of these organisms. Fungal chitosan is more advantageous than crustacean-based chitosan, because the former can be produced in a controlled environment all year round independent of the seasonal seafood industry [8]. Fungal growth and chitosan productivity and properties can be controlled through the manipulation of fermentation and extraction conditions to give chitosan more consistent and desired physico-chemical properties compared with chitosan obtained from crustacean sources [9]. Moreover, the extraction process for isolating chitosan from fungal cell walls is relatively simpler and milder than obtaining chitosan from crustacean shells, thus producing less waste [9]. In terms of chitosan properties, chitosan produced by most of the identified producer fungal strains has roughly the same degree of deacetylation, and relatively lower viscosities and molecular weight than crustacean-derived chitosan, which improves its potential for processing and modification for its numerous applications particularly in agriculture and biomedical engineering [10].

Agro-industrial crops, by-products and residues can be used as inexpensive carbon sources for growing the fungi, which can help alleviate some environmental concerns on the disposal of these waste materials [11]. These materials are particularly suited for use as substrates and microbial growth support in solid-state fermentation (SSF) processes, which exploit the natural growth conditions of filamentous fungi and their capabilities to biosynthesize and excrete hydrolytic enzymes to directly ferment solid biomass substrates. SSF methods have been applied previously for the production of fungal chitosan using a variety of agro-industrial residues and fungal strains, though not as prevalent as using submerged liquid fermentation (SLF) cultures. Potato chip processing wastes [12], cottonseed hulls and corn residues [13], sweet potato [14], and residues from the manufacture of soymilk and mungbean noodles [15] were utilized as fermentation substrates for growth and chitosan production of various filamentous fungal strains. The filamentous fungus Mucor rouxii, a Zygomycete strain, has been well known to produce high amounts of chitosan using SLF [8] [16] [17]. The possibility of utilizing this fungus for SSF of agro-residues for chitosan production has not been tested previously.

Soybean processing residues, such as soybean meal and soy hulls, are excellent options for use as fermentation media/substrates for the growth of fungi that can produce chitosan due to their abundance and high nutrient content. According to the American Soybean Association, one bushel (60 lbs) of soybeans contain on a mass basis approximately 18.3% oil, 80% meal, and 1.7% of other residues such as the hulls and mill runs [18]. With a total annual soybean production of 89.5 million metric tons, the amount of soymeal production is approximately 71.6 million metric tons while that of soybean hulls is estimated to be 1.5 million metric tons [19]. The conventional use of soybean meal includes supplementation of animal feed due to its high carbohydrate and protein content.

However, such use is limited due to the presence of oligosaccharides that are indigestible by non-rumi- nant livestock and poultry [20] . To some extent, soy hulls can be used as animal feed fillers. But due to their relatively low protein and high fiber content, they are not as widely used as soymeal for this application. Instead, they are used as a boiler fuel additive. These characteristics, however, can make both soybean meal and hulls an ideal solid-state fermentation (SSF) substrate for the growth and cultivation of fungi that can be used to produce chitosan. The fungal chitosan product, with all its interesting molecular properties and applications, can be used as a renewable and "green" biobased alternative not only for crustacean chitosan, but also for petroleum based polymers.

The goal of this study is to investigate fungal chitosan production via solid-state fermentation of soybean processing residues using Mucor rouxii. Experiments were conducted to determine the optimum cultivation and extraction conditions for maximum chitosan yield. The physicochemical characteristics of the produced chitosan were then determined using analytical methods.

# EXPERIMENTAL

## Materials

Mucor (Amylomyces) rouxii ATCC 24905, the fungal strain used in this study, was purchased from the American Type Culture Collection (ATCC). The freeze-dried culture was rehydrated with sterile deionized (DI) water and stock cultures were generated by cultivation of the rehydrated fungal cells on potato dextrose agar (PDA) slants at 25°C for five days. The stock cultures were then stored and maintained at 4°C until use. Soybean meal and hulls provided by three local soybean processors was tested as a solid-state fermentation substrate. These were then stored in airtight plastic containers in a refrigerator at 4°C until use. Representative samples of the soybean meal and hulls from the different sources were sent to Northland Labs, Inc. (Northbrook, IL) for analysis of carbohydrates, proteins, moisture, and ash using standard AOAC methods. The soymeal were analyzed and used as received while whole soy hulls were homogenized to granular form using a Waring blender. Composite samples consisting of equal amounts from each of the three suppliers were used for the fermentation process. Commercial grade sodium hydroxide pellets and Optima grade glacial acetic acid were purchased from Fisher Scientific (Pittsburgh, PA, USA). Low molecular weight chitosan purchased from Sigma Aldrich (St. Louis, MO, USA) was used as the reference chitosan for fungal chitosan characterization. The chemicals and standards were used as received.

# Fungal SSF Cultivation and Chitosan Production

The fungal spore suspension inocula used in the SSF cultures were prepared as follows: ten milliliters of sterile DI water was transferred aseptically into a M. rouxii ATCC 24905 stock PDA slant. The slant tube was then gently inverted ten times to suspend the fungal spores in water. After this, 2-mL of the spore suspension was immediately transferred aseptically into a new PDA plate. The spore suspension was distributed onto the PDA surface using a sterile disposable L-spreader. The inoculated plates were incubated at 30°C for five days. After the incubation period, the entire mycelial growth on the agar surface was scraped using a sterile metal spatula and transferred into a 250-mL flask containing 100 mL of sterile DI water. The mixture was gently shaken to suspend the fungal spores in the water. After this, the suspension was filtered through a sterile coarse filter paper into a new sterile collecting flask to remove the large agar and fungal debris. The soybean residue substrates were prepared and inoculated as a single batch in a sterile autoclavable plastic bag. Sterile DI water was added to the filtered fungal spore suspension to attain the desired initial moisture content setting (30%, 50%, or 70% w/w). The pH of the sterile culture dilution water was previously set according to the desired initial pH setting of the substrate (pH 4, 5, or 6). The wet soymeal mash was shaken and mixed thoroughly inside the bag and then distributed in 10-gram portions (wet basis) in sterile polystyrene deep petri dishes. The petri dishes containing the soybean meal fungal cultures were then incubated at the desired temperature setting (25°C, 30°C, or 35°C) for the set cultivation period (3, 6, 9, 12, or 15 days). All experimental treatments were tested in triplicate petri dish cultures.

## Measurement of Fungal Growth

The extent of growth of M. rouxii in the soybean meal cultures was determined indirectly by spectrophotometric measurement of glucosamine in samples of the fermented soymeal substrate according to a method published elsewhere [21].

## Chitosan Extraction

Fungal chitosan produced in the soybean residue cultures was extracted and isolated using an approach based on previously published methods [12] [22] [23]. The entire contents of each petri dish soy residues culture were processed first by extraction of residual proteins and other alkali soluble materials with NaOH solution with the desired concentration (1 M or 46% w/v) and volume-to-solids ratio (30:1 or 10:1, mL/g dry mass). Two different heated extraction methods were also tested: autoclaving at 121°C, 15 psi for 15 min or refluxing at 46°C for 8 h. Alkali insoluble materials (AIMs) were recovered by vacuum filtration and washed to neutral pH with DI water. The washed AIMs were transferred and weighed into tared glass petri dishes and dried in an oven at 60°C overnight or until a constant weight was achieved. The dry weight and moisture content of the AIMs were then calculated using the wet and dry weights measured. Chitosan was extracted from the AIMs with 2% (v/v) aqueous acetic acid. Two different volume-to-solids ratios (40:1 or 10:1, mL/g dry mass AIMs) and heat treatments (autoclaving at 121°C, 15 psi for 15 min or refluxing at 95°C for 8 h) were tested. After the extraction period, the slurries were transferred into tared centrifuge tubes and centrifuged at 3000 RPM for 20 minutes, after which the insoluble fraction was discarded. The pH of the supernatant was adjusted to pH 10 by the dropwise addition of 4 M NaOH in order to precipitate the chitosan out of solution. The suspension was then centrifuged at 3000 RPM for 20 minutes, after which the supernatant was discarded. The chitosan pellets were then washed successively with DI water, 95% ethanol, and acetone; frozen at −20°C overnight; freeze-dried; and weighed to calculate the chitosan yield (% w/w dry mass of initial soymeal substrate).

## Characterization of Fungal Chitosan by FTIR Spectroscopy

The Fourier transfer infrared (FTIR) spectra of the product fungal chitosan from soymeal fermentation were obtained by analyzing solid powder samples with an Alpha-E FTIR spectrometer (Bruker Optics Inc., Billerica, MA, USA). Sixty-four scans were conducted per sample between wavelength number of 4000 - 700 cm$^{-1}$ at a resolution of 2 cm$^{-1}$. The degree of deacetylation (DDA) of the chitosan samples was estimated according to the method of [24]. The absorbance peak at wavelength number ≈ 1560 cm$^{-1}$ (amide II) was used as the characteristic signal while the peak at wavelength number ≈ 1070 cm$^{-1}$ (C-O bond stretching) was used as the reference signal. DDA values were calculated using the absorbance ratio $A_{1560}/A_{1070}$ and the corresponding calibration curve relating the absorbance ratio to DDA.

# RESULTS AND DISCUSSION

## Characterization of Soybean Processing Residues

Following analysis, the soybean meals were found to contain almost 50% by weight proteins, which is roughly five times that of the soy hulls. In terms of the carbohydrate content, soy hulls had around 79% by weight of carbohydrates, which is almost twice that of the carbohydrates in soymeal. Both meal and hulls contain similar residual fats while the meal has slightly higher moisture and ash contents than the hulls. The high carbohydrates and protein content of these residues could serve as carbon and nitrogen sources for direct fungal growth under solid-state cultivation and support the rationale behind the selection of these agricultural residues as fermentation substrates for chitosan production.

## Effect of Substrate Type

Pure soybean meal, soy hulls, and 1:1 meal-hull mixtures were tested as SSF substrates for growth and chitosan production by M. rouxii ATCC 24905. The results (Figure 1) show that fungal growth (as measured by glucosamine in the culture) under solid-state cultivation conditions in a pure soymeal culture or in a 1:1 meal/hull mixture were not significantly different, while growth in pure soy hull was the lowest. This result was expected as the substrates containing the soymeal contained higher protein (nitrogen) contents than pure soy hulls, resulting in the increased biosynthesis of fungal mycelial materials. On the other hand, chitosan yield was highest from the pure soymeal cultures (1.63% ± 0.16% w/w dry substrate) while those from the pure soy hull (1.26% ± 0.15%) and 1:1 meal/hull mixtures (0.97 ± 0.15) were not significantly different. The chitosan yield displayed a positive correlation with fungal growth levels for both the pure soymeal and the 1:1 meal:hull mixtures.

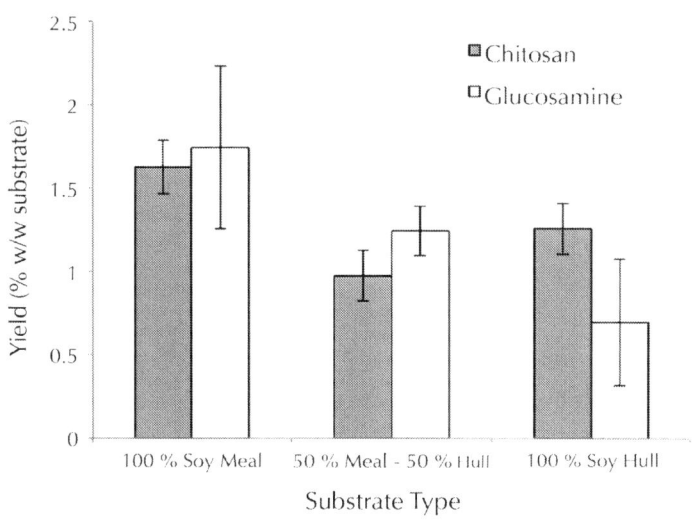

**Figure 1**: Growth and chitosan production by Mucor rouxii ATCC 24905 on different soybean processing residues. Incubation temperature—30°C, initial pH unadjusted, initial mo- isture content—50% (w/w), and cultivation time—six days.

Being a fungal cell wall component, the production of chitosan is considered to be growth-associated (i.e., production rates correlate with growth rate) under carbon- and nutrient-rich condition. However, under a condition of high carbohydrate-to-protein proportion such as in soy hull cultures, the chitosan yield displayed an opposite trend relative to fungal growth between the pure soy hull and 1:1 meal/hull cultures. It appears that despite the lower fungal growth level in the soy hull culture, the fungal cell wall mass generated contained a relatively higher composition of chitosan. Based on these results, pure soymeal substrates were utilized in the succeeding fermentation tests.

## Effect of Incubation Period

The effect of the length of the cultivation period on fungal growth and chitosan yield using pure soymeal SSF cultures is shown in Figure 2. All other fermentation variables were kept constant at the following levels: 50% (w/w) initial substrate moisture content, 30°C incubation temperature, and unadjusted initial substrate pH. The fungal growth pattern (as glucosamine) showed a possible diauxic pattern, with a first stationary period attained between 6 to 12 days of cultivation followed by a sharp increase at day 15. It is possible that during the first stationary period, the fungal cells utilized the readily available adhered carbon and nutrients in the soymeal particle surfaces for partial growth and assembly of hydrolytic enzymes (cellulase, proteases, etc.). During the stationary period, the synthesized enzymes were utilized by the fungal cells for the degradation of the polymeric carbohydrates and proteins into more readily assimilable carbon and nutrients, which were then made available for a succeeding exponential growth phase. Despite the observed increase in growth, the results show that maximum chitosan yield was achieved after only six days of cultivation after which chitosan yields fluctuated (not significantly different). Previous studies have indicated that during the attainment of the stationary growth phase, the fungal cell walls may have begun to lose their fluidity and potential for chitosan formation due to consolidation by chitin crystallization and bond formation with other cell wall components, which could lead to reduced extraction yields [22] [25] . Thus, six days was considered to be the optimum length of cultivation period for maximum chitosan yield and was maintained in the succeeding experiments.

## *Effect of Initial Moisture Content*

Figure 3 shows the effect of the initial substrate moisture content on growth and chitosan yield of M. rouxii ATCC 24905 in SSF soymeal cultures. In these experiments, the incubation period was set at six days based on the optimum level obtained in the prior experiments, the incubation temperature at 30°C, and initial substrate pH was not adjusted. The lowest moisture level tested (30% by wt.) resulted in the lowest fungal biomass growth and chitosan production. However, fungal growth and chitosan yields were not significantly different between the 50% and 70% initial moisture content treatments (chitosan yield is between 1% and 2% w/w dry substrate).

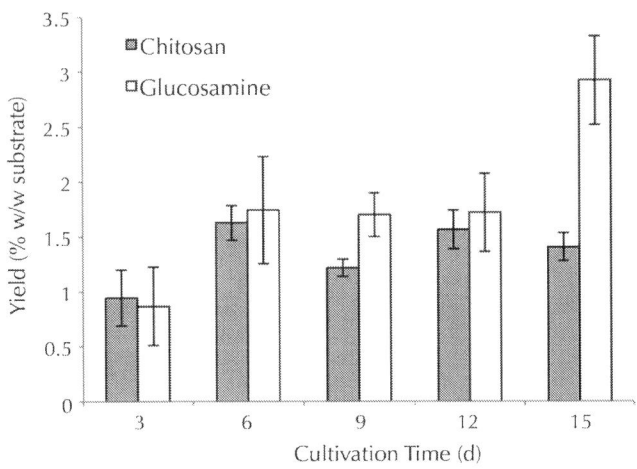

**Figure 2**: Effect of cultivation time on growth and chitosan yield by Mucor rouxii ATCC 24905 on pure soymeal substrate. Incubation temperature—30°C, initial pH unadjusted, initial moisture content—50% (w/w), and cultivation time—six days.

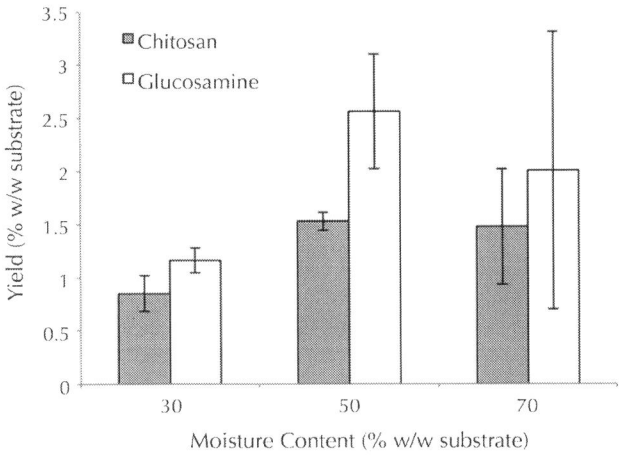

**Figure 3**: Effect of initial substrate moisture content on growth and chitosan production by Mucor rouxii ATCC 24905 on pure soymeal substrate. Incubation temperature—30°C, initial pH unadjusted, and cultivation time—six days.

The moisture content of the substrate is a critical parameter related to the water activity of the microorganisms in SSF processes where moisture levels are relatively low. The fungal cells need an appropriate level of water in which to exercise the various biochemical reactions for the enzymatic breakdown and uptake of the biomass substrate. The results in this study indicated that M. rouxii ATCC 24905 was able to tolerate a lower substrate moisture content of 50% for growth and chitosan production in SS unlike other fungal strains previously studied for chitosan production via SSF, that were grown in 60% - 70% moisture substrates [12] [22] [26] . Using less water to moisten the soybean meal substrate could mean less dilution of nutrients and growth factors due to water addition in the culture. Furthermore, using soymeal cultures with 70% initial moisture content appeared to produce high variability in the results. Thus, the initial soymeal substrate moisture content of 50% was considered to be the optimum level for maximizing fungal growth and chitosan production using the strain M. rouxii ATCC 24905.

## Effect of Incubation Temperature

Figure 4 shows the effect of incubation temperature on fungal growth and chitosan production in solid-state cultures of M. rouxii ATCC 24905.

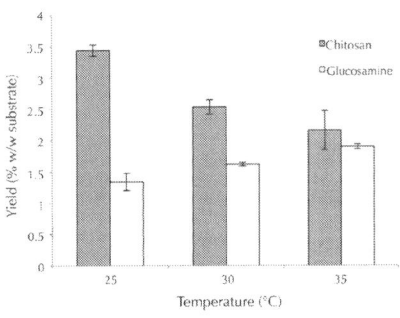

**Figure 4**: Effect of incubation temperature on growth and chitosan production by Mucor rouxii ATCC 24905 on pure soymeal substrate. Ini-

tial pH unadjusted, initial moisture content—50% (w/w), and cultivation time—six days.

In these experiments, the incubation period was set at six days and the initial moisture content of the soymeal substrate was 50% (w/w) based on the optimum levels obtained in the prior experiments. The initial substrate pH was not adjusted. The experiments demonstrated opposite trends of growth and chitosan yield with increasing incubation temperature. The highest chitosan yield of around 3.44% ± 0.09% (w/w of total solid substrate) was achieved at 25°C and decreased considerably at incubation temperatures of 30 and 35°C. On the other hand, total fungal growth was shown to increase as the incubation temperature increased. A higher incubation temperature appeared to favor fungal growth in the solid soymeal culture but results in lower chitosan yields. A possible explanation is that the fungal cell walls might be losing their fluidity in favor of a denser and more consolidated structure dominated by chitin, which results in reduction of chitosan composition and/or extractability at incubation temperatures higher than 25°C. This could be a defense mechanism by the fungal cells to counteract the potentially detrimental effects of high temperatures on fungal growth and metabolism. On the operational standpoint, incubation at lower temperatures could lead to lower costs attributed to heating and humidification of fermentation chambers but could imply the necessary fermentation bed cooling mechanisms to counteract the buildup of temperature gradients across the substrate bed thickness.

## *Effect of Initial Substrate pH*

Figure 5 shows the effect of the initial pH of the moistened soybean meal substrate on fungal growth and chitosan production in solid-state cultures of M. rouxii ATCC 24905. In these experiments, the incubation period was set at six days, the incubation temperature at 30°C, and initial substrate moisture content was at 50% (w/w) according to the results of the prior experiments. The results obtained demonstrate that there are no significant differences in terms of fungal growth and chitosan yields among the initial substrate pH levels investigated. It seems logical to select the unadjusted pH as the best condition from the operational cost standpoint, which would not require prior treatment of the soymeal substrate with acid and/or alkali for pH adjustment. However, an initial pH setting between 5 and 6 might be ideal for helping to minimize the variability of the product chitosan yields and characteristics. Additional tests will be conducted to confirm this hypothesis.

## *Effect of Extraction Conditions*

In addition to testing the effect of the fermentation variables on fungal growth and chitosan yield, different extraction methods were also tested to determine their effects on chitosan recoveries. Different methods were selected from literature. The results of these tests are summarized inFigure 6. In these experiments, the fermentation conditions were kept constant at six days cultivation period, 50% initial substrate moisture content, unad- justed initial substrate pH, and 30°C incubation temperature.

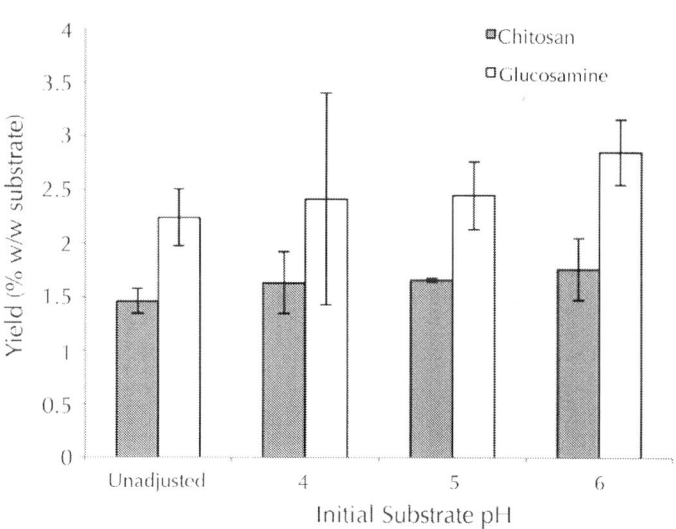

**Figure 5**: Effect of initial substrate pH on growth and chitosan production by Mucor rouxii ATCC 24905 on pure soymeal substrate. Incubation

temperature? 30°C, initial moisture content—50% (w/w), and cultivation time—six days.

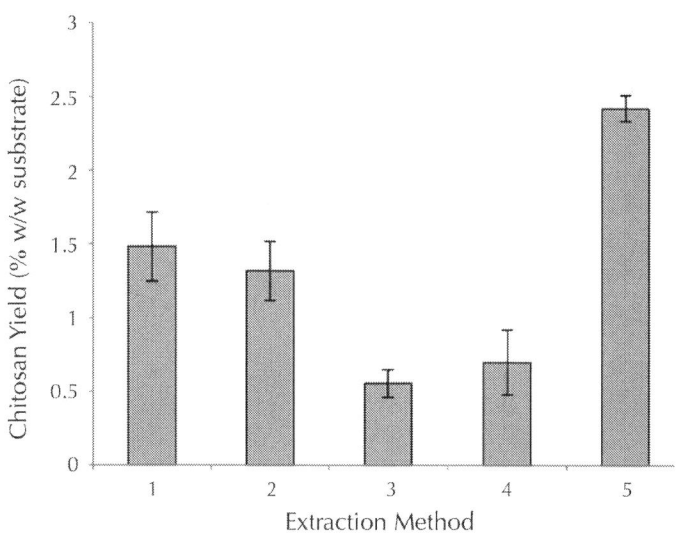

**Figure 6**: Effect of extraction method on growth and chitosan yield by Mucor rouxii ATCC 24905. Method 1—1 M NaOH at 121°C for 20 min (30 mL per g substrate) followed by 2% (v/v) acetic acid (40 mL per g substrate) at 95°C with refluxing for 8 h. Method 2—1 M NaOH at 121°C for 20 min (10 mL per g substrate) followed by 2% (v/v) acetic acid (10 mL per g substrate) at 95°C with refluxing for 8 h. Method 3—46% (w/w) NaOH at 46°C for 13 h (30 mL per g substrate) followed by 2% (v/v) acetic acid (40 mL per g substrate) at 95°C with refluxing for 8 h. Method 4—46% (w/w) NaOH at 46°C for 13 h (10 mL per g substrate) followed by 2% (v/v) acetic acid (10 mL per g substrate) at 95°C with refluxing for 8 h. Method 5—1 M NaOH at 121°C for 20 min (30 mL per g substrate) followed by 2 % (v/v) acetic acid (40 mL per g substrate) at 121°C for 20 min.

Figure 6 shows that chitosan recoveries differed significantly among the different methods investigated with regards to the concentration of reagents used as well as the severity of the heat treatment method used. For instance, Methods 1 and 2 utilized the same reagent concentrations but with differing volume ratios, while Method 2 used a lower volume of reagent. Although a higher volume-to-mass ratio for both the base and acid reagents (Method 1) resulted in a slightly higher chitosan recovery, the results were not significantly different. The same case was observed with Methods 3 and 4, which used a higher base concentration but with heating and refluxing instead of autoclaving. Compared with Methods 1 and 2, Methods 3 and 4 resulted in a significantly lower chitosan recovery. This finding implies the importance of the base-assisted deproteinization step for maximizing the chitosan yield. However, Figure 5 also shows that when a more severe reaction condition (i.e., autoclaving) was employed for the acetic acid extraction of chitosan step, the chitosan recovery was significantly higher than all the other extraction methods tested. Despite this, a more thorough economic and cost-benefit study should be made to justify the use of more severe extraction conditions and correlations must be made with the effect of fermentation conditions.

# Characterization of Fungal Chitosan Using FTIR

The chitosan extracted from Mucor rouxii ATCC 24905 fungal cultures in soybean meal was characterized by FTIR spectroscopy and compared with commercially available chitosan. The purpose of this analysis was to confirm the identity of the extracted chitosan pellets and to estimate the degree of deacetylation (DDA) of the fungal chitosan product, an important parameter that influences the activity of the chitosan polymer for its intended applications. The DDA of the fungal chitosan produced from solid-state fermentation of soybean meal was estimated using the absorbances of the FTIR signals at approximate wavelength numbers of 1560 and 1070 cm$^{-1}$ (Shigemasa et al., 1996). As shown in Figure 7, the signal at 1560 cm$^{-1}$ corresponds to the characteristic band of the amide group (Amide II) in the N-acetylglucosamine residue of chitin, which accounts for more than 90% of the monomer residues in the polymer chain. Although definitions vary among different sources, the chitin chain is identified as chitosan if the polymer chain contains a significant proportion of the deacetylated glucosamine residue. Thus the intensity of the band at 1560 cm$^{-1}$ is expected to decrease as the DDA increases. The signal at 1070 cm$^{-1}$ corresponding to C-O bond stretching in the chitin/chitosan monomers was used as the reference band. Using the calibration curve developed by Shigemasa et al., the DDA of representative samples of fungal chitosan from different fermentation and extraction conditions was estimated to be between 55% - 60%.

## CONCLUSIONS

Solid-state fermentation (SSF) of soybean processing residues using the fungus Mucor rouxii ATCC 24905 for chitosan production was for the first time investigated. The optimum SSF conditions for maximum chitosan yields were determined.

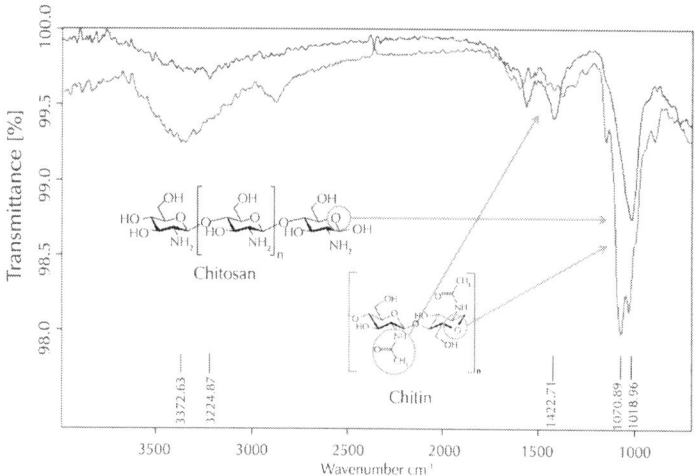

**Figure 7**: FTIR spectra of commercial chitosan (red) and Mucor rouxii ATCC 24905 chitosan from fermentationofpure soymeal substrate (blue).

FTIR analysis of the fungal chitosan derived from soybean meal fermentation confirmed its molecular identity and estimated its degree of deacetylation to be around 55% - 60%. Extraction conditions such as the concentration of acid and base reagents and severity of heat treatment were also shown to influence the chitosan yield.

# ACKNOWLEDGEMENTS

The authors would like to acknowledge the United Soybean Board (USB) for providing the funding for this research (USB Project # 1440-512-5283): Archer Daniels Midland, Zeeland Farms, and CHS for supplying the soybean meal and hull samples; and Dr. Linda Kim-Habermehl of USB/Omni Tech International, Ltd. for her valuable inputs on the project.

# REFERENCES

1. Castro, S.P.M. and Paulín, E.G.L. (2012) Is Chitosan a New Panacea? Areas of Application. In: Karunaratne, D.N., Ed., The Complex World of Polysaccharides, ISBN: 978-953-51-0819-1, InTech. http://www.intechopen.com/books/the-complex-world-of-polysaccharides/is-chitosan-a-new-panacea-areas-of-application
2. Kaur, S., and Dhillon, G.S. (2014) The Versatile Biopolymer Chitosan: Potential Sources, Evaluation of Extraction Methods and Applications. Critical Reviews in Microbiology, 2, 155-175. http://dx.doi.org/10.3109/1040841X.2013.770385
3. Badawy, M.E.I. and Rabea, E.I. (2011) A Biopolymer Chitosan and Its Derivatives as Promising Antimicrobial Agents against Plant Pathogens and Their Applications in Crop Protection. International Journal of Carbohydrate Chemistry, 2011, Article ID: 460381, 29 p. http://dx.doi.org/10.1155/2011/460381
4. Dutta, P.K., Dutta, J., and Tripathi, V.S. (2004) Chitin and Chitosan: Chemistry, Properties and Applications. Journal of Scientific & Industrial Research, 63, 20-31. http://nopr.niscair.res.in/handle/123456789/5397

5. Bruck, W.M., Slater, J.W., and Carney, B.F. (2011) Chitin and Chitosan from Marine Organisms. In: Kim, S.K., Ed., Chitin, Chitosan, Oligosaccharides, and Their Derivatives: Biological Activities and Applications, CRC Press, Boca Raton, 11-24.
6. Knezevic-Jugovic, Z., Petronijevic, Z., and Smelcerovic, A. (2011) Chitin and Chitosan from Microorganisms. In: Kim, S.K., Ed., Chitin, Chitosan, Oligosaccharides, and Their Derivatives: Biological Activities and Applications, CRC Press, Boca Raton, 25-36.
7. Peter, M.G. (2002) Chitin and Chitosan in Fungi. Biopolymers Online, 6, 123-157.http://dx.doi.org/10.1002/3527600035. bpol6005
8. White, S.A., Farina, P.R. and Fulton, I. (1979) Production and Isolation of Chitosan from Mucor rouxii. Applied and Environmental Microbiology, 38, 323-328.
9. Tan, S.C., Tan, T.K., Wong, S.M. and Khorb, E. (1996) The Chitosan Yield of Zygomycetes at Their Optimum Harvesting Time. Carbohydrate Polymers, 30, 239-242.http://dx.doi. org/10.1016/S0144-8617(96)00052-5
10. Pochanavanich, P. and Suntornsuk, W. (2002) Fungal Chitosan Production and Its Characterization. Letters in Applied Microbiology, 35, 17-21.http://dx.doi.org/10.1046/j.1472-765X.2002.01118.x
11. McGahren, W.J., Perkinson, G.A., Growich, J.A., Leese, R.A. and Ellestad, G.A. (1984) Chitosan by Fermentation. Process Biochemistry, 19, 88-90.
12. Kleekayai, T. and Suntornsuk, W. (2010) Production and Characterization of Chitosan Obtained from Rhizopus oryzae Grown on Potato Chip Processing Waste. World Journal of Microbiology and Biotechnology, 27, 1145-1154. http:// dx.doi.org/10.1007/s11274-010-0561-x
13. Wang, W., Du, Y., Qiu, Y., Wang, X., Hu, Y., Yang, J., Cai, J. and Kennedy, J.F. (2008) A New Green Technology for Direct Production of Low Molecular Weight Chitosan. Carbohydrate

14. Polymers, 74, 127-132. http://dx.doi.org/10.1016/j.carbpol.2008.01.025
14. Nwe, N. and Stevens, W.F. (2002) Production of Fungal Chitosan by Solid Substrate Fermentation Followed by Enzymatic Extraction. Biotechnology Letters, 24, 131-134. http://dx.doi.org/10.1023/A:1013850621734
15. Suntornsuk, W., Pochanavanich, P. and Suntornsuk, L. (2002) Fungal Chitosan Production on Food Processing By-Products. Process Biochemistry, 37, 727-729.http://dx.doi.org/10.1016/S0032-9592(01)00265-5
16. Bartnicki-Garcia, S. and Nickerson, W.J. (1962) Isolation, Composition, and Structure of Cell Walls of Filamentous and Yeast-Like Forms of Mucor rouxii. Biochimica et Biophysica Acta, 58, 102-119. http://dx.doi.org/10.1016/0006-3002(62)90822-3
17. Chatterjee, S., Adhya, M., Guha, A.K. and Chatterjee, B.P. (2005) Chitosan from Mucor rouxii: Production and Physico-Chemical Characterization. Process Biochemistry, 40, 395-400. http://dx.doi.org/10.1016/j.procbio.2004.01.025
18. American Soybean Association (2014) SoyStats 2014. Soybean Facts.http://soystats.com/facts/
19. American Soybean Association (2014) SoyStats 2014. US Yield and Production: Production History. http://soystats.com/u-s-yield-production-production-distory/
20. Zhang, L.Y., Li, D.F. and Qiao, S.Y. (2003) Effects of Stachyose on Performance, Diarrhea Incidence and Intestinal Bacteria in Weanling Pigs. Archives of Animal Nutrition, 57, 1-10. http://dx.doi.org/10.1080/0003942031000086662
21. Roopesh, K., Ramachandran, S., Nampoothiri, K.M., Szakacs, G. and Pandey, A. (2006) Comparison of Phytase Production on Wheat Bran and Oilcakes in Solid-State Fermentation by Mucor racemosus. Bioresource Technology, 97, 506-511. http://dx.doi.org/10.1016/j.biortech.2005.02.046
22. Crestini, C., Kovac, B. and Giovannozzi-Sermanni, G. (1996) Production and Isolation of Chitosan by Submerged and Solid-

State Fermentation from Lentinus edodes. Biotechnology and Bioengineering, 50, 207-210.http://dx.doi.org/10.1002/bit.260500202

23. Rane, K.D. and Hoover, D.G. (1993) Production of Chitosan by Fungi. Food Biotechnology, 7, 11-33. http://dx.doi.org/10.1080/08905439309549843

24. Shigemasa, Y., Matsuura, H., Sashiwa, H. and Saimoto, H. (1996) Evaluation of Different Absorbance Ratios from Infrared Spectroscopy for Analyzing the Degree of Deacetylation in Chitin. International Journal of Biological Macromolecules, 18, 237-242.http://dx.doi.org/10.1016/0141-8130(95)0179-3

25. Nwe, N., Furuike, T. and Tamura, H. (2010) Production of Fungal Chitosan by Enzymatic Method and Applications in Plant Tissue Culture and Tissue Engineering: 11 Years of Our Progress, Present Situation and Future Prospects. In: Elnashar, M., Ed., Biopolymers, ISBN: 978-953-307-109-1, InTech. http://dx.doi.org/10.5772/10261http://www.intechopen.com/books/biopolymers/production-of-fungal-chitosan-by-enzymatic-method-and-applications-in-plant-tissue-culture-and-tissu

26. Khalaf, S.A. (2004) Production and Characterization of Fungal Chitosan under Solid-State Fermentation Conditions. International Journal of Agriculture & Biology, 6, 1033-1036. http://www.fspublishers.org/published_papers/4088_..pdf

# Chapter 5

# A Promising Material by Using Residue Waste from Bisphenol a Manufacturing to Prepare Fluid-Loss-Control Additive in Oil Well Drilling Fluid

Zhi-Lei Zhang[1], Feng-Shan Zhou[1], Yi-He Zhang[1], Hong-Wei Huang[1], Ji-Wu Shang[1], Li Yu[1], Hong-Zhen Wang[2], and Wang-Shu Tong[1]

[1]National Laboratory of Mineral Materials, School of Materials Science and Technology, China University of Geosciences, Beijing 100083, China

[2]Institute of Chemistry, Chinese Academy of Sciences, Beijing 100190, China

# ABSTRACT

The residues mixture from Bisphenol A manufacturing process was analyzed. Fourier transform infrared (FTIR) spectroscopy, gas chromatography-mass spectrometry (GC-MS), and nuclear magnetic resonance (NMR) were used to characterize the residues. The results indicated that the residues were complex mixture of several molecules. 3-(2-Hydroxyphenyl)-1,1,3-trimethyl-2,3-dihydro-1H-inden-5-ol and phenol were the main components of the residues. The technical feasibility of using it as phenol replacement in fluid-loss-control additive production was also investigated. The fluid-loss-control capacity of the novel additive was systematically investigated. It was discovered that the well fluid-loss performance of the prepared additive can be achieved, especially at high temperature.

# INTRODUCTION

Bisphenol A (BPA) is an organic compound with two phenolic hydroxyl groups [1]. Generally, it is synthesized by condensation of phenol with acetone [2–4]. For reactions involving the substitution of a proton in an aromatic ring, both the rate of reaction and the equilibrium distribution of products are influenced by the density of electrons at the centre of reaction [5]. However, due to the high reactivity of the system, many byproducts can be produced and are present in the reaction mixture. For example, a crude product stream consisted of 41% BPA, 36.2% ortho,para-isomer, 1.1% ortho,ortho-isomer, 14.2% phenol, 3.5% chromane, 0.05% flavan, and 12% of unidentified compounds [6]. The formulas of some by-products are shown in Figure1. After numerous and diverse purification processes, the impurities (such as the excess phenol, cyclic dimmers derived from BPA, chroman-based compounds, spirobiindane compounds, and the like) were separated from BPA and finally formed the complex residue compound.

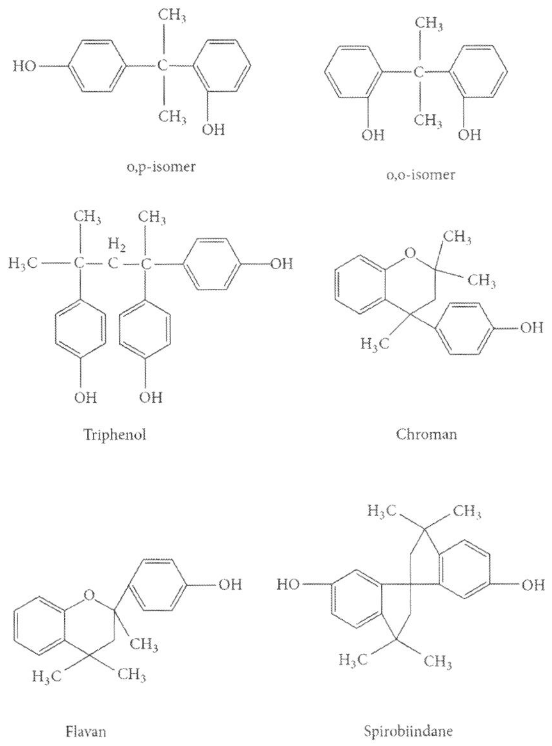

**Figure 1:** Formulas of the impurities.

Being an important industrial chemical, BPA has been widely used [7, 8], and the demand for BPA production has increased [1]. With the growing production of BPA [9], there will be more and more residues from the BPA industry. It is predicted that there will be over $50 \times 10^3$ ton residues produced annually in China. According to European Waste Catalogue (EWC), the residues are absolute hazardous (07 01 08*). So, appropriate disposal and recycling of the residues are necessary for environmental protection and public health. The residues are generally burned as a means of disposal. Because the incineration technology is not qualified; treatment and recycling of the residues in China have been a thorny problem. The previous researches mainly concentrated on the treatment of the by-products and recovery of the useful phenol [10–12]. Hence, it is imperative to develop methods to utilize the industry waste. So far, there have only been relatively limited researches in this respect. The residues have been used in furan no-bake foundry binders and phenolic-based foundry shell resin formulations [13].

The petroleum drilling fluids, more commonly known as drilling muds, are complex chemical systems necessary for oil development. Among other functions, a drilling mud needs to create a thin low-permeability cake that protects permeable production formations. Fluid-loss-control additives form filter cakes surrounding the well bore to retard the loss of drilling fluid into permeable formations [14–17]. It has been common practice in the oil well drilling industry to employ starches, starch derivatives, cellulose derivatives, and water-soluble gums to reduce the filtrate volume of water base muds. Although these materials reduce the fluid loss of drilling muds, they are not thermally stable. Generally, water-soluble sulfonated phenolic condensate is obtained from the reaction of phenol, formaldehyde, and anhydrous sodium sulfite [18]. As a drilling mud additive, it has good fluid-loss-control performance and resistance to high temperature [19]. However, when the treated muds are subjected to extreme thermal environments, such products should be further modified to enhance the properties. And with raw material prices, the commercial pressure of competition is also growing. Utilization of waste materials for modification or substitution of primary resources has been a widespread concern [20–22]. Phenol has an ortho-para directing activation group, hydroxyl. It could be condensed and sulfonated to produce a fluid-loss-control additive under the presence of formaldehyde and sodium sulfite. Although there are many complicate components coexisting in the residues, most of them have the activation group hydroxyl and could react like phenol. The condensation polymerization by using this kind of residues is more likely a copolycondensation. The resultant products will have the same essential functional group as the products from condensation and sulfonation by using phenol. So, it is feasible by using the residues to displace phenol as a raw material to produce fluid-loss-control additive.

The purpose of this study was to confirm the components of the residue mixture and evaluate the technical feasibility of using it as phenol replacement in fluid-loss-control additive production.

# EXPERIMENTAL DETAILS

## Materials

The residues mixtures were obtained from Nanjing Zunyu Chemical Co., Ltd., China. It was obvious that the waste material was shiny black solid (Figure 2(a)). It was highly brittle, and the powder (Figure 2(b)) could be obtained after grinding. And then, the powder was stirring under the refluxing temperature in distilled water and sodium hydroxide (NaOH) solution for an hour, respectively. In Figure 2(c), the dispersion of the powder was markedly different. In distilled water, the powder was gathering into a mass slowly, but it dispersed well in the NaOH solution. The raw bentonite was obtained from Dagang Oilfield (Group) Co., Ltd., Tianjin, China. The sulfonated methyl phenolic resin obtained from the Third Exploration Company of Tarim Petroleum Exploration and Development Headquarters was marked as $M_1$ and obtained from Derun Chemical Co., Ltd., Binzhou, China, and was marked as $M_2$. Sulfonated lignite obtained from Hairong Industry & Trade Co., Ltd., Korla, China, and Beijing Institute of Exploration Engineering was marked as $M_3$ and $M_4$, respectively. All the chemicals such as phenol, formaldehyde, sodium sulfite, sodium bisulfite, sodium hydroxide, sodium chloride, acetone, and cyclohexane were analytical grade and obtained from Beijing Chemical Plant, China. A homemade column chromatograph with a length and width of 40 cm and 3 mm, respectively, packed with silica gel (ZCX.H, 200–300 mesh size, Branch of Qingdao Haiyang Chemical Plant, China) was used in the study. Standard HSGF254 HPTLC plates (100 × 25 mm) were purchased from Yantai Chemical Industry Research Institute, China.

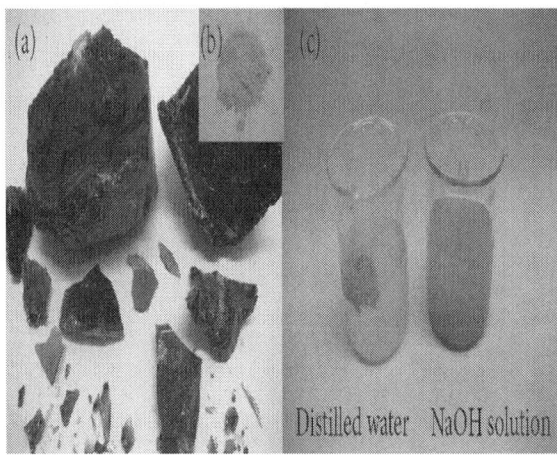

**Figure 2:** Comparison between the residues before (a) and after (b) grinding and the dispersion in distilled water and NaOH solution (c).

## Methods

FTIR was conducted on the residues on a Perkin-Elmer SP100 FTIR spectrometer by the KBr disc technique. It was operated in the ATR (attenuated total reflectance) mode, and 32 scans were collected at a resolution of 4 cm$^{-1}$.

An Agilent Technologies 7890A gas chromatograph equipped with the Agilent 5975C mass spectrometer (Agilent 7890A/5975C GC-MS System) was used to identify the organic compositions. The system was operated at 40°C for 1 min and then up to 300°C at a ramping rate of 10°C/min and held for 5 min. One μL of the samples was injected in split mode with a split ratio of 10:1. An Agilent J&W column, with 30 m in length, 0.25 mm internal diameter, and 0.25 μm film thicknesses was adopted in the separation system. High-purity helium was the carrier gas and introduced at a constant rate of 1.4 mL/min.

The eluent used in the column chromatography experiment was an acetone/cyclohexane (1/5, v/v) mixture which was determined by previous thin-layer chromatography experiments. The materials obtained by column chromatography were dried at 50°C for 24 h in a vacuum oven to a constant weight. Nuclear magnetic resonance spectra ($^1$H-NMR, 400 MHz) were acquired on a Bruker-400 spectrometer with the 1 mm TXI micro liter probe using deuterated chloroform as the solvent.

The fresh-water base mud containing 50 g/L of sodium bentonite was prepared by mixing the raw bentonite and fresh water at a certain ratio, stirring for 15 min at a high speed of 10,000 rpm and aging for 24 h at room temperature. The testing mud was obtained with an addition of fluid-loss-control agents and/or sodium chloride into the base mud. The fluid-loss-control properties of the muds, including API filtrate volume (i.e., $FL_{API}$) and high-temperature/high-pressure filtrate volume (i.e., $FL_{HTHP}$), were determined according to American Petroleum Institute (API) specifications and Chinese SY/T5621-93 specifications, respectively. In this test process, a ZNS-2A medium-pressure and GGS42-2 high-pressure filtration apparatus (made by Qingdao Haitongda Special Instruments Co., Ltd., China) were used. Figure 3 shows the schematic of the filtration apparatus. The rheological parameters, such as apparent viscosity (AV), plastic viscosity (PV), and yield point (YP), were determined by a ZNN-D6 rotating viscometer.

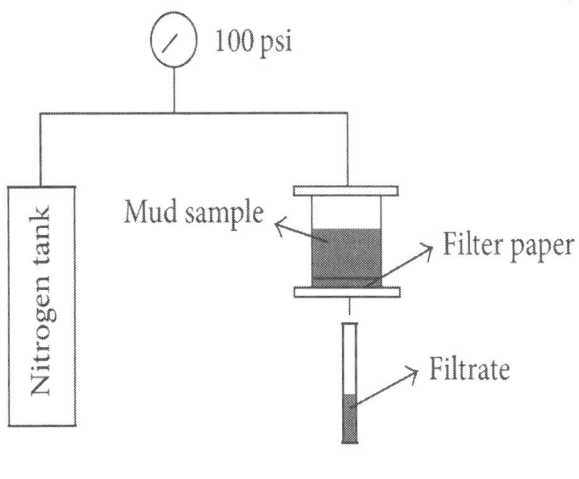

(a)

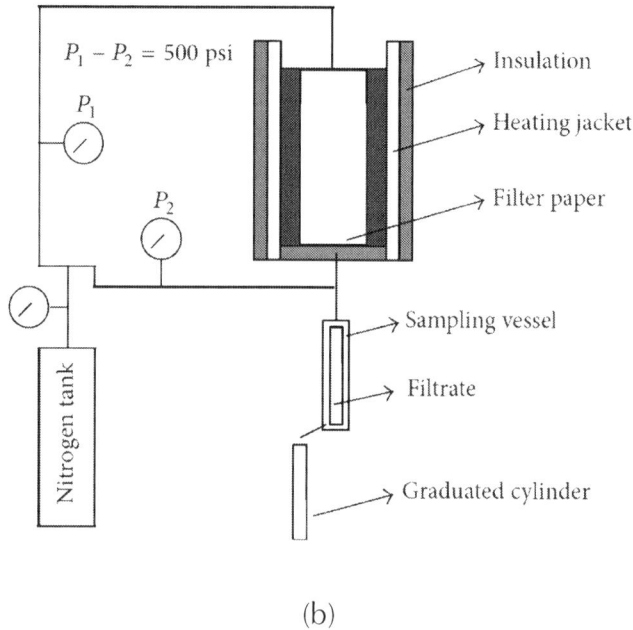

(b)

**Figure 3:** Schematic of the API (a) and HTHP (b) filtration test apparatus.

# Preparation of Fluid-Loss-Control Additive RPF

A 3-neck flask with a stirrer, thermometer, and reflux condenser was charged with the residues, formaldehyde water solution (37% w/w), sodium hydroxide, and water. The mixture was heated slowly to a certain temperature. After stirring for about an hour, a second portion of formaldehyde water solution, sodium hydroxide and water was added into the mixture with sodium sulfite and sodium bisulfite. After synthesis, the production was pan-dried at 70°C and then crushed into powders over 80 mesh sieve for use in mud tests. The nomenclature of this fluid-loss-control additive is designated as RPF here. And the PF was prepared using phenol to replace the residues. Furthermore, there are very little by-products produced from this process which is environmentally green.

## Orthogonal Array Experimental Design

In the present study, an L25 (56) orthogonal array (OA) was used to investigate the effect of formulas and parameters. The orthogonal array was listed in Table 1, and the data analysis was carried out through the range analysis. After the orthogonal experiments and subsequent data analysis, the magnitudes were reflected.

**Table 1**: Levels and factors affecting the fluid-loss properties

| Levels | Factors | | | | | |
|---|---|---|---|---|---|---|
| | A (°C)[a] | B[b] | C[c] | D[d] | E (mL)[e] | F(h)[f] |
| 1 | 60 | 2.77 | 17.12 | 3.61 | 29.7 | 3 |
| 2 | 70 | 1.39 | 5.71 | 2.71 | 39.6 | 4 |
| 3 | 80 | 0.93 | 3.42 | 1.80 | 49.5 | 5 |
| 4 | 90 | 0.69 | 2.45 | 1.20 | 59.4 | 6 |
| 5 | 100 | 0.56 | 1.90 | 0.90 | 69.3 | 7 |

[a]A: reaction temperature.
[b]B: mass ratio of residue to formaldehyde.
[c]C: mass ratio of residue to catalyst.
[d]D: mass ratio of residue to sulfonating agent.
[e]E: amount of water.
[f]F: reaction time.

# RESULTS AND DISCUSSION

## Spectroscopy Analysis

The infrared spectrum of the residues with the characteristic hydroxyl group stretching band at 3424 cm$^{-1}$ can be observed in Figure 4. The hydroxyl group band in the residues and RPF does not as strong as that in the PF. This could be attributed to the complex components of the residues and RPF. The absorption peaks at 2962 cm$^{-1}$ and 1468 cm$^{-1}$ can be assigned to the vibrational modes of methyl and/or methylene. The peaks at 1605 cm$^{-1}$ corresponded to the carbon-carbon double bond stretching vibration of the benzene ring skeleton. An observable sulfonate group antisymmetric stretching vibration band at 1195 cm$^{-1}$ and a carbon-sulfur bond stretching band at 1042 cm$^{-1}$ can be found in the spectrum of RPF and PF but not in the spectrum of the residues. This feature is considered to be characteristic of the sulfonation reaction.

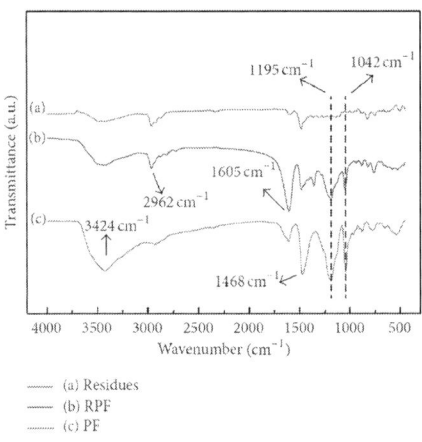

**Figure 4:** FTIR spectra of the residues, RPF, and PF.

Gas chromatography-mass spectrometry (GC-MS) is used to isolate and determine the organic molecules. The chromatogram of residues is depicted in Figure 5(a). Separation and peak profiles are accomplished in 30 min and several components can be isolated. The component contents (with the retention time of 6.8 min, 8.8 min, 14.0 min, 21.5 min, and 22.9 min) are 34.5%, 7.6%, 5.6%, 24.6%, and 6.4%, respectively (Table 2). Figures 5(b),5(c), 5(d), 5(e), and 5(f) show the partial mass spectra of these five constituents marked as 1$^\#$, 2$^\#$, 3$^\#$, 4$^\#$, and 5$^\#$. The molecular ions in the two main components, 1$^\#$ and 2$^\#$, have m/z values of 94 and 268, respectively, suggesting that number 1 is phenol, number 2 is 3-(2-hydroxyphenyl)-1,1,3-trimethyl-2,3-dihydro-1H-inden-5-ol. Phenol has already been used to produce phenolic resin and sulfonated phenolic resin. Number 2 has two phenolic hydroxyl groups, so it could condense with formaldehyde as well.

**Table 2:** GC/MS results of the residues

| Peak | Retention time (min) | Assignments | Molecular ion | Content (%) |
| --- | --- | --- | --- | --- |
| 1# | 6.8 | Phenol | 94.0 | 34.5 |
| 2# | 21.5 | A[a] | 268.0 | 24.6 |
| 3# | 8.8 | Undecane | 156.0 | 7.6 |
| 4# | 14.0 | B[b] | 174.0 | 5.6 |
| 5# | 22.9 | C[c] | 268.0 | 6.4 |

[a]A: 3-(2-hydroxyphenyl)-1,1,3-trimethyl-2,3-dihydro-1H-inden-5-ol

[b]B: 1,5,7-trimethyl-1,2,3,4-tetrahydronaphthalene.

[c]C: 3-(4-hydroxyphenyl)-1,1,3-trimethyl-2,3-dihydro-1H-inden-5-ol.

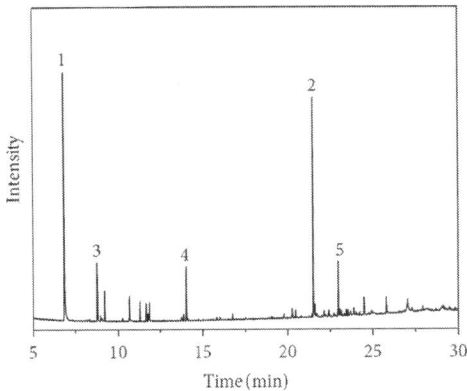

(a)

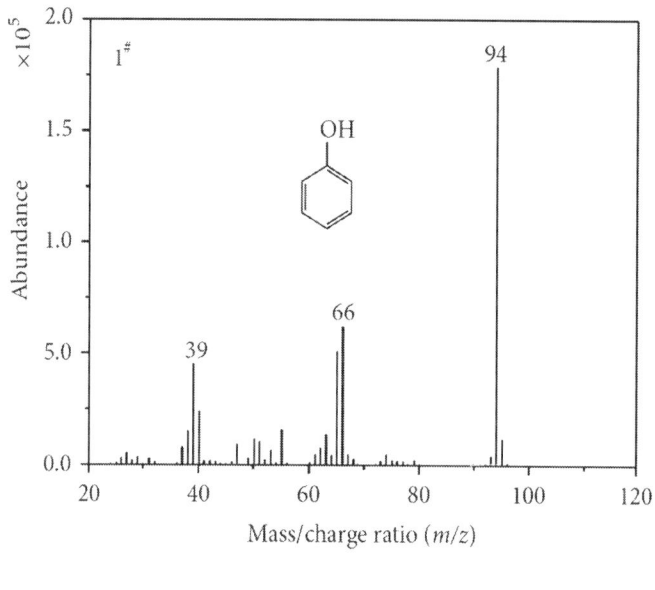

(b)

(c)

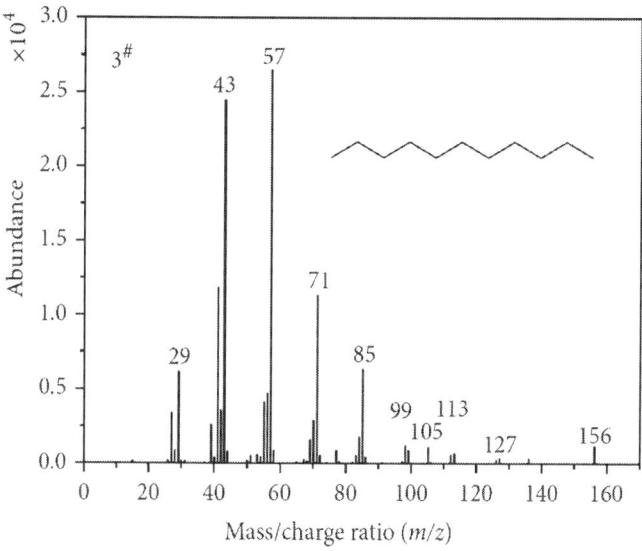

(d)

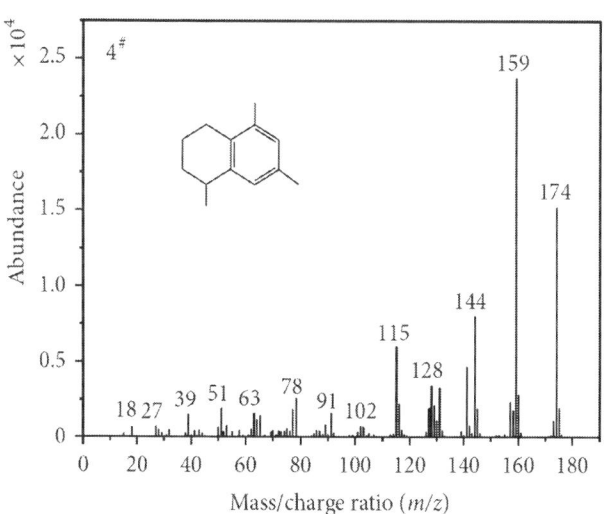

(e)

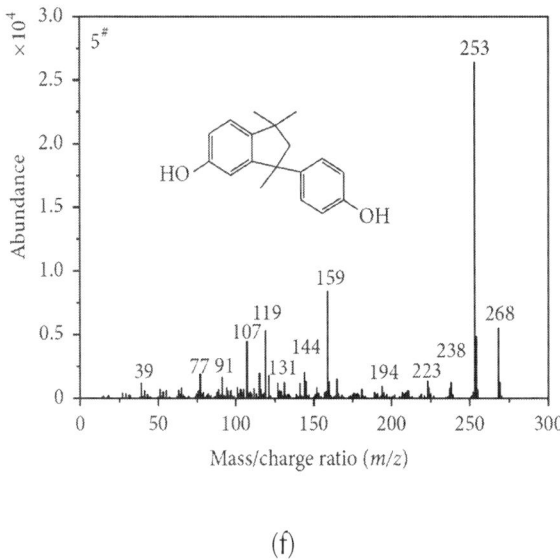

(f)

**Figure 5:** Chromatogram of the residues (a) and mass spectra of the five components from the residues (peaks 1–5 in the chromatogram) at different retention times of (b) 6.8 min, (c) 21.5 min, (d) 8.8 min, (e) 14.0 min, and (f) 22.9 min.

There are many components in the residues. Some of the components have little content, similar structures, and polarity. Hence, it is difficult to separate the components absolutely. In our column chromatography experiments, there are five components obtained and the two main components are further identified by $^1$H NMR and the results are in agreement with those from GC-MS. As shown in Figure 6(a), the signal at 5.051 ppm corresponds to the hydroxyl group (1). In the region of 6.8–7.3 ppm, there are signals assigned to the aromatic protons (2, 3, and 4). Hence, compounds number 1 is indeed phenol, also known as carbolic acid. In Figure 6(b), the resonance at 5.388 ppm corresponds to the hydroxyl group (1). The signals at 1.619 ppm, 0.858 ppm, and 2.346 ppm arise from protons 2, 3, and 4, respectively. In the region of 6.683–7.267 ppm, there are signals assigned to the aromatic protons and deuterated chloroform. Thus, number 2 is identified to be 3-(2-hydroxyphenyl)-1,1,3-trimethyl-2,3-dihydro-1H-inden-5-ol.

# A Promising Material by Using Residue Waste from Bisphenol ...   133

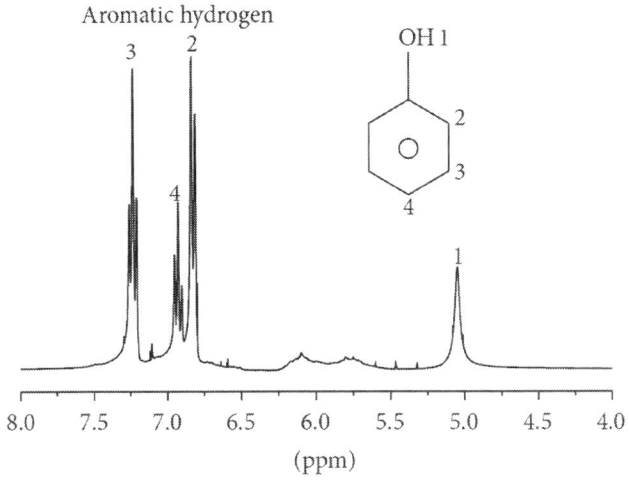

(a)

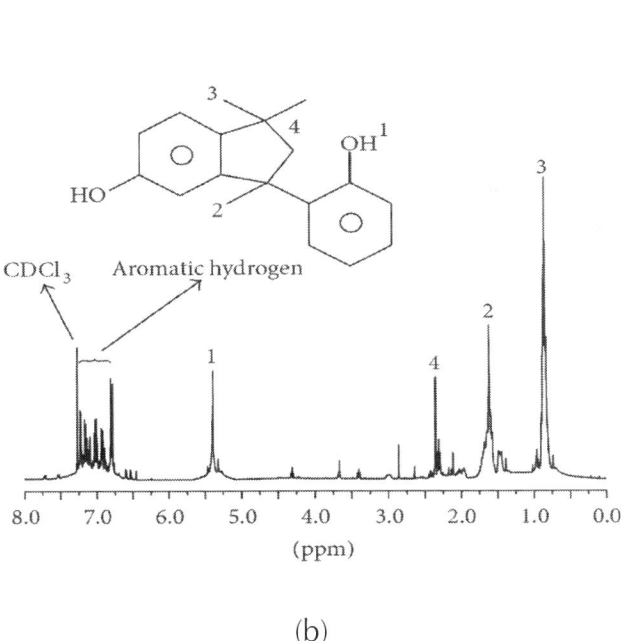

(b)

**Figure 6:** Proton nuclear magnetic resonance spectrum acquired from the two main components of the residues.

Hydroxyl is an ortho-para directing group. Due to the reactive positions, these two components could be condensed and sulfonated under the presence of formaldehyde and sodium sulfite. The simple schematic diagrams of the production were shown in Figure 7; (a) was PF and (b) was RPF. This is not the real structure but a possible structure that imaged to illustrating the condensation polymerization and the sulfonation reaction. The real structure of the reaction products will be more complex. The other components do not play a major role in this study because of the relatively tiny amount or the absence of active functional group.

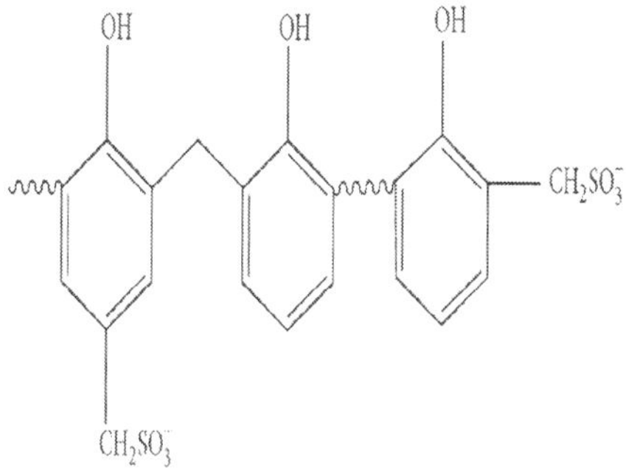

(a)

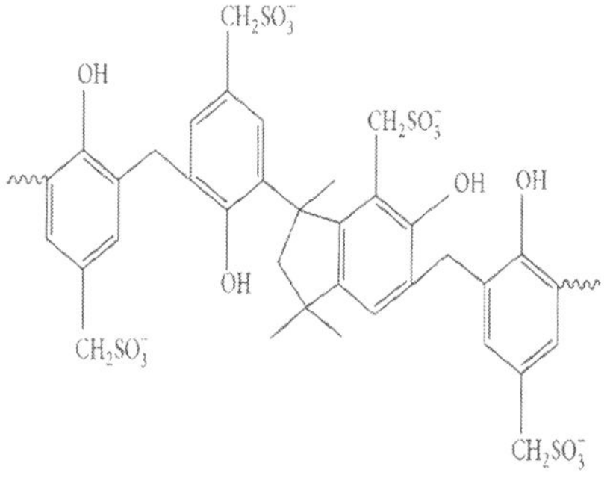

(b)

**Figure 7:** Structural representation of PF and RPF.

## Effect of Formulas and Parameters on Filtration Properties

There are two important parameters in range analysis: $K_{ij}$ and $R_i$. $K_{ij}$ is defined as the sum of the evaluation indexes of all levels (j=1,2,3,4) in each factor (i=A,B,C,D). $K'_{ij}$ (mean value of $K_{ij}$) is used to determine the optimal level and the optimal combination of factors. The optimal level for each factor could be obtained when $K'_{ij}$ is the largest. $R_i$ is defined as the range between the maximum and minimum value of $K'_{ij}$ and is used for evaluating the importance of the factors [23]. According to the OA25 matrix, twenty-five experiments were carried out and their filtrate volume results were shown in Table 3. This table shows that the range of $FL_{API}$ was from 10 mL to 30 mL and the range of $FL_{HTHP}$ was from 50 mL to 138 mL.

The mean values of K( $K'_{ij}$ ) for different factors at different levels in the range analysis were shown in Table 4. As mentioned, for each factor, the higher mean value ( $K'_{ij}$ ) indicates the larger effect on filtration loss. As shown in Table 4, for API filtrate properties, $K'_{ij}$ was the lowest at these combinations $A_4B_4C_2D_5E_3F_1$, and, for HTHP properties, $K'_{ij}$ was the lowest at $A_1B_2C_5D_3E_4F_1$. Compared with the range values of different factors ($R_i$), the factors' levels of significance are as follows: for API, C (6.10) > B (4.20), F (4.20) > A (3.70) > D (3.48) > E (3.12); for HTHP,A (31.20), B (31.20) > E (27.60) > C (26.00) > D (19.60) > F (16.00).

**Table 3:** $FL_{API}$ and $FL_{HTHP}$ in OA25 matrix

| Trial number | Factors | | | | | | Results | |
|---|---|---|---|---|---|---|---|---|
| | A | B | C | D | E | F | $FL_{API}$(mL) | $FL_{HTHP}$(mL) |
| 1 | 1 | 1 | 1 | 1 | 1 | 1 | 22 | 78 |
| 2 | 1 | 2 | 2 | 2 | 2 | 2 | 23 | 72 |
| 3 | 1 | 3 | 3 | 3 | 3 | 3 | 24 | 60 |

| | | | | | | | | |
|---|---|---|---|---|---|---|---|---|
| 4 | 1 | 4 | 4 | 4 | 4 | 4 | 20 | 72 |
| 5 | 1 | 5 | 5 | 5 | 5 | 5 | 20 | 90 |
| 6 | 2 | 1 | 2 | 3 | 4 | 5 | 19 | 72 |
| 7 | 2 | 2 | 3 | 4 | 5 | 1 | 27 | 92 |
| 8 | 2 | 3 | 4 | 5 | 1 | 2 | 22 | 100 |
| 9 | 2 | 4 | 5 | 1 | 2 | 3 | 25 | 126 |
| 10 | 2 | 5 | 1 | 2 | 3 | 4 | 30 | 138 |
| 11 | 3 | 1 | 3 | 5 | 2 | 4 | 26 | 110 |
| 12 | 3 | 2 | 4 | 1 | 3 | 5 | 23 | 104 |
| 13 | 3 | 3 | 5 | 2 | 4 | 1 | 18 | 50 |
| 14 | 3 | 4 | 1 | 3 | 5 | 2 | 26 | 124 |
| 15 | 3 | 5 | 2 | 4 | 1 | 3 | 26 | 116 |
| 16 | 4 | 1 | 4 | 2 | 5 | 3 | 20 | 74 |
| 17 | 4 | 2 | 5 | 3 | 1 | 4 | 22 | 56 |
| 18 | 4 | 3 | 1 | 4 | 2 | 5 | 26 | 104 |
| 19 | 4 | 4 | 2 | 5 | 3 | 1 | 10 | 86 |
| 20 | 4 | 5 | 3 | 1 | 4 | 3 | 27 | 108 |
| 21 | 5 | 1 | 5 | 4 | 3 | 2 | 21 | 60 |
| 22 | 5 | 2 | 1 | 5 | 4 | 3 | 25 | 68 |
| 23 | 5 | 3 | 2 | 1 | 5 | 4 | 20 | 90 |
| 24 | 5 | 4 | 3 | 2 | 1 | 5 | 25 | 112 |
| 25 | 5 | 5 | 4 | 3 | 2 | 1 | 24 | 96 |

**Table 4:** Range analysis data of the $FL_{API}$ and $FL_{HTHP}$

| Value name | A | B | C | D | E | F |
|---|---|---|---|---|---|---|
| $FL_{API}$ | | | | | | |
| $K'_1$ | 21.80 | 21.78 | 25.80 | 23.40 | 23.40 | 20.30 |

| | | | | | | |
|---|---|---|---|---|---|---|
| $K'_2$ | 24.70 | 24.00 | 19.70 | 23.30 | 24.80 | 23.10 |
| $K'_3$ | 23.90 | 22.10 | 25.80 | 23.10 | 21.68 | 24.50 |
| $K'_4$ | 21.00 | 21.20 | 21.80 | 24.08 | 22.00 | 23.60 |
| $K'_5$ | 23.08 | 25.40 | 21.38 | 20.60 | 22.60 | 22.70 |
| R | 3.70 | 4.20 | 6.10 | 3.48 | 3.12 | 4.20 |
| $FL_{HTHP}$ | | | | | | |
| $K'_1$ | 74.40 | 78.80 | 102.40 | 101.20 | 92.40 | 80.40 |
| $K'_2$ | 105.60 | 78.40 | 87.20 | 89.20 | 101.60 | 89.00 |
| $K'_3$ | 100.80 | 80.80 | 96.40 | 81.60 | 89.60 | 92.00 |
| $K'_4$ | 85.60 | 104.00 | 89.20 | 88.80 | 74.00 | 93.20 |
| $K'_5$ | 85.20 | 109.60 | 76.40 | 90.80 | 94.00 | 96.40 |
| R | 31.20 | 31.20 | 26.00 | 19.60 | 27.60 | 16.00 |

## Filtration Properties

According to the optimal conditions shown above, we have changed and modified the experiment conditions and parameters and finally prepared the fluid-loss-control additive RPF and PF. As a fluid-loss-control additive, RPF and PF could improve the properties of drilling fluid. Series of fresh-water and salt-water mud formulations were prepared, and the filtrate volumes were measured. To probe into the effect of dosage of the fluid-loss-control additive on filtration properties, test of API filtration, and HTHP filtration were performed. From Figures 8 and 9, it can be observed that the fluid-loss volumes decreased with more additives used. These additives dispersed into the base mud and then could be adsorbed onto the clay surface under the action of hydrogen bond between the hydroxyl and the clay surface [24]. So, there exists a saturated adsorption amount. When the additive addition was over dose, the excess additives have little influence on filtration properties. As shown in Figures 8 and 9, when the concentration of the additive was higher than 20 g/L, the fluid-loss volumes decreased slowly.

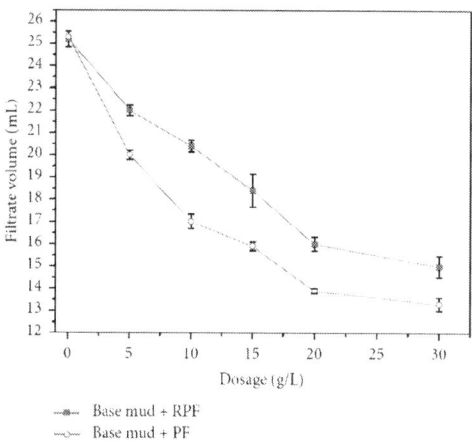

**Figure 8:** The effect of RPF and PF concentration on the API filtration loss of fresh water mud.

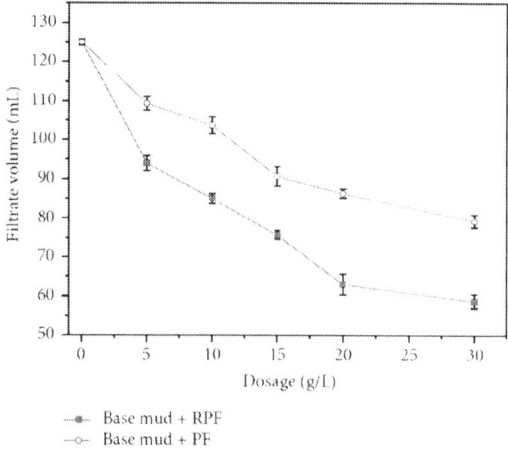

**Figure 9:** The effect of RPF and PF concentration on the HTHP filtration loss of fresh water mud.

From the analysis shown in Figures 5 and 6 and the possible structure of the product shown in Figure 7, it was expected that RPF would have better HTHP filtrate properties than PF. Segmental motion of molecular chain is dependence on temperature. The higher the temperature is, the more intense the segmental motion is. When the segmental motion is intense enough, molecular repulsion will be stronger than attraction and the molecular chain breaks. This was called degradation. There are two situations when polymer degrades; the molecular weight of the additive decreases and the amount of functional group of the additive decreases. These situations would cause the fluid-loss-control performance deterioration. Polymer will degrade at high temperatures, but the degree of the degradation is different because of the different molecular structure. If there are more aromatic groups in the molecular chain, the chain segments are bigger, and the temperature needs to be higher to make the segments motion. 3-(2-Hydroxyphenyl)-1,1,3-trimethyl-2,3-dihydro-1H-inden-5-ol and phenol are the main components in the residues. Phenol condensation polymer is endowed well HTHP fluid-loss-control performance by introducing benzene ring into the molecular structure. 3-(2-Hydroxyphenyl)-1,1,3-trimethyl-2,3-dihydro-1H-inden-5-ol has the activation group hydroxyl, and the molecular structure is more rigid than phenol. So, the resultant additives will have well HTHP fluid-loss-control performance by using these residues.

There are two competitive factors that affect the fluid-loss-control performance of RPF. (1) The components in the residue are complex, and some of them do not have the essential reactive group. These components cannot participate in the reaction and have no effect to improve the HTHP performance. (2) Most of the components in the residue are aromatic derivative, and they have the same reactive group like phenol. These components have the essential functional group as fluid-loss-control additives. Meanwhile, the side chain of the RPF molecular is bigger by introducing these aromatic derivatives into the condensation reaction as the constitutional unit than introducing phenol. The bigger side chain provides RPF better thermostability.

It can be seen in Figure 9 that the HTHP fluid-loss-control performance of RPF is better than PF. This can be explained as, in the fresh water mud, the thermo stability of the molecular chain is more important for the HTHP performance than the amount of the functional groups. The positive effect of the rigid molecular structure of the aromatic derivatives is more than the negative effect of the inactive components. This is why fewer components in residue could result in better HTHP fluid-loss-control performance in fresh water mud of RPF than that of PF.

To show the stability of RPF as a fluid-loss-control additive in saline solutions, filtration loss test of 20 g/L additive in 50 g/L bentonite mud with 40 g/L NaCl was performed. It is important to have additive for drilling mud that are stable in saline environments. Without additive, bentonite mud would lose its fluid-loss-control properties. According to the DLVO theory, it is thought to be explained as that the electrolyte solutions will compress the electrostatic double layer, affect hydration and coagulation stability of clays, and lead to the dehydration of the additives. Therefore, salt resistance additives should have enough adsorbing groups to connect with the clay surface and enough hydration groups to control the free water, and the hydration groups should be insensitive to electrolyte. The RPF have phenolic hydroxyl groups as the adsorbing groups and sulfonate groups as the hydration groups which have been proved to have well salt resistance [25]. RPF was expected to have salt resistance like PF. From the experiments data shown in Figures 10 and 11, unlike the expected results, RPF took a relatively poor effect compared with PF. It might be caused by the less active ingredient content. In the presence of salt solutions, the content of sulfonate groups plays the major role of the salt resistance. There are some inactive components in the residues, which could not participate in the sulfonation reaction. However, from the experiment results shown in Figure 9, it is obvious that although the effective component is less, the HTHP performance is still better.

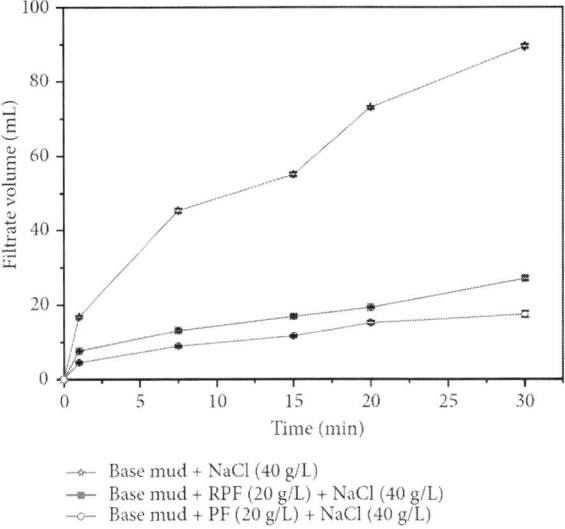

**Figure 10:** API filtration loss of RPF and PF in base mud and NaCl (40 g/L) solutions.

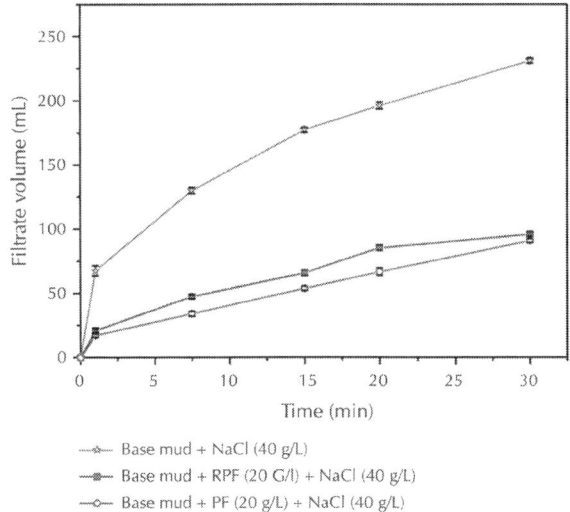

**Figure 11:** HTHP filtration loss of RPF and PF in base mud and NaCl (40 g/L) solutions.

For comparison, the properties of RPF and some commercial fluid-loss-control additives were shown in Table 5. It can be seen that the API filtrate volume of RPF is 16 mL and the HTHP filtrate volume of RPF is 62 mL. Through the comparison, it reveals that the HTHP performance of RPF is better than these commercial products. The viscosity of RPF is higher, and the yield point is in the middle.

**Table 5:** Properties of RPF compared to commercial fluid-loss-control additives

| Fluid-loss-control additive[a] | AV (mPa·s) | PV (mPa·s) | YP (Pa) | $FL_{API}$ (mL) | $FL_{HTHP}$ (mL) |
|---|---|---|---|---|---|
| RPF | 21.5 | 12.0 | 9.5 | 16.0 | 62.0 |
| $M_1$ | 20.5 | 7.0 | 13.5 | 16.0 | 88.0 |
| $M_2$ | 17.5 | 7.0 | 10.5 | 10.0 | 64.0 |
| $M_3$ | 13.5 | 9.0 | 4.5 | 11.0 | 66.0 |
| $M_4$ | 14.0 | 8.0 | 6.0 | 12.0 | 72.0 |

[a] The dosage of fluid-loss-control additive was 20 g/L.

## CONCLUSIONS

The residue wastes produced from the BPA industry were complex compounds. Phenol and 3-(2-hydroxyphenyl)-1,1,3-trimethyl-2,3-dihydro-1H-inden-5-ol were the two main components of the residues. The fluid-loss-control additive produced using the residues showed better HTHP fluid-loss-control performance than some commercial products. This study indicates that the residues have potential application in fluid-loss-control additive manufacture. Using it to substitute phenol will improve the high temperature performance and endow the additive with significant cost advantage.

# DISCLAIMES

The paper is authors owned work, is original and unpublished, and is not being considered for publication elsewhere.

# ACKNOWLEDGMENTS

This work is jointly supported by the Special fund of co-construction of Beijing Education Committee and the Fundamental Research Funds for the Central Universities no. 2652013062.

# REFERENCES

1. W. T. Tsai, "Human health risk on environmental exposure to Bisphenol-A: a review," Journal of Environmental Science and Health, vol. 24, no. 2, pp. 225–255, 2006.
2. P. Michele and C. Giuseppe, Braz. Pedido PI BR, 7, 903, 483, 1979.
3. V. N. Sheemol, I. R. Unni, and C. Gopinathan, "Catalysis by heteropoly acids: formation of bisphenol A from phenol and acetone," Indian Journal of Chemical Technology, vol. 8, no. 4, pp. 298–300, 2001.
4. A. E. Donald, B. R. Lawrence, C. Ye-mon, and J. R. Lawrence, Upflow Fixed Bed Reactor With Packing Elements, WO Patent, 1997.
5. C. D. Nenitescu, Chimie Organica, vol. 2, p. 47, 1980.
6. Z. N. Verkhovskaya, M. Ya. Klimenko, L. B. Vystavkina et al., "Composition of the by-products of diphenylolpropane synthesis and their recovery," Neftpererab Nefekhim, vol. 5, pp. 34–35, 1973.
7. A. V. Krishnan, P. Stathis, S. F. Permuth, L. Tokes, and D. Feldman, "Bisphenol-A: an estrogenic substance is released from polycarbonate flasks during autoclaving," Endocrinology, vol. 132, no. 6, pp. 2279–2286, 1993.

8. C. A. Staples, P. B. Dorn, G. M. Klecka, S. T. O›Block, and L. R. Harris, "A review of the environmental fate, effects, and exposures of bisphenol A," Chemosphere, vol. 36, no. 10, pp. 2149–2173, 1998.
9. F. Jiao, X. Sun, and Z. Pang, "Production and market analysis of Bisphenol A," Chemistry & Industry, vol. 26, pp. 21–33, 2008.
10. S. Evitt, C. Chi, M. S. Lee, and D. Palmer, Process For Recovering Phenol From A BPA Waste Stream, 2010.
11. J. C. Carnahan, Phenol Recovery From Bisphenol-A Waste Streams, 1981.
12. S. J. Shafer, J. Pressman, and J. L. Lee, Method For Recovering Material Values From Bisphenol Tars, 2001.
13. K. K. Chang, M. C. Clingerman, M. L. Lott, and J. T. Schneider, Use of Bisphenol A tar in furan No-bake Foundry Binders, 1999.
14. P. L. Moore, Drilling Practices Manual, The Petroleum Publishing, Tulsa, Oklahoma, 1974.
15. C. L. William and J. P. Gary, Standard Handbook of Petroleum & Natural Gas Engineeringed, Gulf Professional Publishing, Houston, Tex, USA, 2nd edition, 1996.
16. A. D. Patel, E. Stamatkis, and E. Davids, U.S. Patent 6, 247, 543, 2001.
17. J. McDermott, Drilling Mud and Fluid Additives, Noyes Data Corp, London, UK, 1973.
18. K. C. Hsu and Y. F. Lee, "Water-soluble sulfonated phenolic resins. I. Synthesis," Journal of Applied Polymer Science, vol. 57, pp. 1419–1537, 1995.
19. J. J. M. Nahm and D. A. Rowe, U.S. Patent 3, 956, 140, 1976.
20. S. Rukzon and P. Chindaprasirt, "Utilization of bagasse ash in high-strength concrete," Materials and Design, vol. 34, pp. 45–50, 2012.

21. W. Wang and G. Huang, "Characterisation and utilization of natural coconut fibres composites,"Materials and Design, vol. 30, no. 7, pp. 2741–2744, 2009. ·
22. H. Y. Aruntas, M. Gürüb, M. Dayı, and T. İlker, "Utilization of waste marble dust as an additive in cement production," Materials & Design, vol. 31, no. 8, pp. 4039–4042, 2010.
23. C. Chuanwen, S. Feng, L. Yuguo, and W. Shuyun, "Orthogonal analysis for perovskite structure microwave dielectric ceramic thin films fabricated by the RF magnetron-sputtering method," Journal of Materials Science, vol. 21, no. 4, pp. 349–354, 2010.
24. D. J. Greenland, "Adsorption of polyvinyl alcohols by montmorillonite," Journal of Colloid Science, vol. 18, no. 7, pp. 647–664, 1963.
25. U. P. Strauss and Y. P. Leung, "Volume changes as a criterion for site binding of counterions by polyelectrolytes," Journal of the American Chemical Society, vol. 87, no. 7, pp. 1476–1480, 1965.

# Chapter 6

# Evaluation of Additives as Corrosion Inhibitors/Antioxidants for High Quality Nano Emulsifiable Oils of Metalworking Fluids

Noura El Mehbad

Faculty of Science, Najran University, Najran, KSA

## ABSTRACT

Surfactants used for formulation metal working fluids. In the present paper, the inhibitive effect of synthesized anionic surfactant P-decyloxy p-sodium sulphonateazobenzene with chemical structure $H_{21}C_{10}$O-Ph-N=N-Ph-$SO_3$Na and 1-ethyl-1-dodecyl-2-sulphonate-4-(hydroxyl ethyl)-piprazine on the corrosion of carbon steel in sulphuric acid solution is measured by the weight loss method.

The relationships between the concentrations of these inhibitors and their surface properties, thermodynamic properties, surface coverage and inhibiting efficiency, have been investigated. The results indicate that P-decyloxy p-sodium sulphonate azobenzene is superior to 1-ethyl-1-dodecyl-2-sulphonate-4-(hydroxyl ethyl)-piprazine at high acid concentration. These inhibitors blend with coupling agent to produce water miscible cutting fluid. Values of surface tension of these additives were measured in water phase and consequently CMC was determined for all inhibitors. The efficiency of these inhibitors depends on their chemical structure and the presence of hydrophilic group to association of the surfactant with metal surface, hence very good wetting properties. The anti-corrosion characteristic of surfactants increases with increasing polar linkages in the aromatic moieties forming chelated and coordinated layer. This is due to the increase in number of sites to chemisorption on the metal surface. These studies have led to much clear evidence of the intimate relationship between the chemical structure of the surfactants and their efficiency. More confirmation for suggested mechanism was investigated by measuring the area occupied per molecule of the surfactant at aqueous phase. The results indicate that the amphoteric surfactant more efficient than anionic surfactant. The antioxidants activities of different dosages of inhibitors were evaluated and suggested mechanism according to their thermodynamics parameters. The oxidation of the oil has been carried for different time intervals. The degradation of the oil has been monitored by total acid formation.

# INTRODUCTION

Metalworking can be defined as processes of modification of the shape of a metal piece. In all the metalworking processes, a considerable amount of heat is evolved due to the internal friction and the friction between tool and metal. Straight oils are used as rolling fluids for rolling of thin sheets of ferrous and nonferrous metals. Straight oils contain base oils and additives. Paraffinic oils are preferred. Also these additives include corrosion inhibitors.

Recently, the mixture of anionic/nonionic polymeric surfactants used as additives for metaworking fluid was studied by Omar 2004 [1].

Most commercial acid inhibitors are organic ammonium salts, organic amine derivatives, organic phosphates, succinimides and imidazoline derivatives [2]-[4]. The inhibitive effects of benzyl triethanol ammonium chloride and its ethoxylated derivative on the corrosion of carbon steel in sulphuric acid were previously investigated [2]. Four cationic corrosion inhibitors having the general formula $ClRN^+H_2(CH_2)_nN^+H_2RCl$ where R is a hydrogen atom or benzyl group, n = 8 or 10 were applied in sulphuric acid solution at temperature up to 78°C. It was found that the alkylation processes improved with surface and thermodynamic properties of the surfactant molecules, and consequently more inhibition effects were obtained [3]. The inhibiting effect of cationic surfactant N,N,N-dimethyl 4-methylbenzyl dodecyl ammonium chloride on mild steel in hydrochloric acid solution was investigated by surface and thermodynamic measurements, weight loss tests, polarization measurements and EDS techniques. It found that, the corrosion inhibition efficiency of the surfactant compound increases with increasing concentration and reached a maximum value at near the critical micelle concentration (CMC). The inhibitor acts through adsorption phenomenon and formation of barrier film [4]. Some amphoteric surfactants N-Decyl-N-benzyl-N-methylglycine(AB) and N-Dodecyl-N-benzyl-N-methylglycine(CD) were evaluated by author. The physicochemical chemical characteristics were investigated. Surface properties, in particular the critical micelle concentration (CMC), the maximum surface excess ($_{CMC}$) and the minimum surface area ($A_{MIN}$) were measured. It is found the surface and thermodynamic properties of the prepared surfactants depend on their hydrocarbon chain length. Also it is found that there is a good relation between surface properties of the additive and their efficiency in depressing the pour point. The mechanism of the depressants action has been suggested according the adsorption of each additive [5]. In our previous work, the synthesis of new additives as pour point and antioxidants was described [6] [7].

The author study new antioxidant for lube oil. This antioxidant dibenzyl s-phenyl thioglyconitrile and other derivatives were prepared by phase transfere catalysts. These compounds were added to oil in different concentrations. The antioxidants activities of different dosages were evaluated and suggested mechanism according to micelle and its thermodynamic. Novel method of inhibiting oxidation was proposed by author. The mechanism was suggested according to surface activity of additive in oil phase. More confirmation for suggested mechanism was investigated by measuring the area occupied per molecule of additive at oil phase. There is a good relationship between the structure of hydrophilic group of the additive and its efficiency. The antioxidants activities of different dosages were evaluated and suggested mechanism according to micelle and its thermodynamics. The oxidation of the oil has been carried for different acid formation. Oxidation stability of lube oil was largely affected by sulphur and aromatic hydrocarbons concentration in oil, with increased sulphur content increase oxidation stability time intervals. Polyalkylphenol formaldehyde sulphonate and its ethoxylate were synthesized and evaluate as pour point depressant, viscosity improver and antioxidant. The efficiency of these additives depends on their chemical structure and degree. The modification of the lyophobic and lyophilic groups, in the structure of the surfactant, may become necessary to maintain surface activity at a suitable level.

Therefore, it is very important to choose the correct surfactants and optimize its concentration to get full efficiency benefit of using base stock.

The purpose of the present work study of prepared P-decyloxy p-sodium sulphonate azobenzene with chemical structure $H_{21}C_{10}$O-Ph-N=N-Ph-$SO_3$Na and 1-ethyl-1-dodecyl-2-sulphonate-4-(hydroxyl ethyl)-piprazine as additives in metalworking straight oil formulations for anti-corrosion and anti-oxidant. These additives differ in head group of surfactant. The author will suggest the field of action mechanism of the additive according to its surface properties.

# EXPERIMENTAL

## Formulation of Metal Working Straight Oil

Base paraffinic oil represents 82% with the additive (5%), other coupling agent about 3% of dodecyl alcohol and sodium oleate about 7% [8].

## Synthesis of the Additives

The following surfactants were prepared and kindly supplied by Omar [9]. Sodium salt of sulphonic acid (0.24) in 100 ml of ethanol was added to 10 ml of concentrated hydrochloric acid. The reaction contents were refluxed for 5 h. Solvents were removed using a rotary evaporator and the residue was dissolved in 100 ml of 1:1 aqueous acetone and 10 ml of concentrated hydrochloric acid, followed by 50 ml of sodium nitrate (0.3 mol) in an icebath. The diazonium salt solution obtained was added to 100 ml of an aqueous solution of 0.37 mol of phenol, 0.35 mol of NaOH and 0.57 mol of sosium carbonate and the mixture was stirred for 12 h at room temperature. Then it was neutrralized with acetic acid and the product was recrystallized to give an orange powder of p-hydroxy—p sodium sulphonate azobenzene with purity of 96% [9]. Equimolar amounts of p-hydroxy—p sodium sulphonate azobenzen decyl bromide was added to sodium ethoxide solution simultaneously. After refluxing for 4 h, the reaction mixture was poured into 500 ml of ice-water and the resulting product was extracted with chloroform. After removal of chloroform using a rotary evaporator, the product was recrystallized from ethanol to give an orange powder of $H_{21}C_{10}$-Ph-N=N-Ph-$SO_3$Na.

The second surfactant 1-ethyl-1-dodecyl-2-sulphonate-4-(hydroxyl ethyl)-piprazine was kindly supplied by omar and prepared according to Reference [10].

## Surface and Interfacial Tension Measurement

Surface tension of different concentrations for $10^{-7}$ to 0.1 mol/L of the synthesized additives was measured by using Kruss Model 8451 in $H_2O$ at 30°C - 50°C according to omar et al. [11].

The physicochemical properties of the base oil are listed in the following Table 1 [8].

A carbon steel electrode was used to evaluate the corrosion rate in sulphoric acid medium by the weight loss method. The chemical composition of the steel is (wt%): C, 0.06 - 0.2; Mn, 0.27 - 0.60; P, 0.048, S, 0.056, and the surface is 3 cm × 3 cm. Before use the electrode is polished successively with emery paper down to 600 grades, degreased with ethyl alcohol, rinesed with doubly distilled water and then weighed.

At optimum conditions the standard test method (ASTM D 4627) was used to evaluate anticorrosion properties of cutting oils. The electrode was placed in baker containing filter paper and diluted metal working fluid in the prescience of the above concentration of sulphoric acid. The best conditions were selected to investigate their surface of samples to determine the corrosion mechanism using scanning electron microscope.

## Oxidation Stability Study

The oxidation test was carried out at 120°C according to ASTM D 943 standard methods. The base stock sample was subjected to oxidation with pure oxygen at a flow rate of 0.1 L/hour for maximum 70 hours.

# RESULTS AND DISCUSSIONS

## Surface and Thermodynamic Properties

Surface tension of two inhibitors has been measured at different temperatures and for different concentrations. Their values are shown in Table 2. The critical micelle concentration CMC was determined at each temperature, the surface excess concentration $\Pi_{max} \times 10^{-11}$ mol/cm$^{-2}$, area occupied per molecule $A_{min}$ nm$^2$, free energy of micellization molecule $G^0$mic KJ/mol and free energy of adsorption at solution/air interface $G^0$ad KJ/mol have been calculated. The results indicate that the CMC of the inhibitors decrease with increasing temperature, while the values of CMC of inhibitors BAM is smaller than ANS. The minimum area per molecule at the aqueous solution/air interface of BAM is slightly bigger than ANS and slightly increases with increasing temperature, this mean that BAM will increase degree of surface coverage of metal surface. The surface excess concentration decreases with increasing temperature for two inhibitors, while BAM prefers adsorption rather than the other. Studying the results in Table 2 shows that the inhibitors BAM has larger values of effectiveness $PC_{20}$, efficiency $_{CMC}$, maximum surface excess $\Pi_{max}$ and minimum surface area $A_{min}$, indicating that this inhibitor is most efficient one that favors adsorption at the air/solution interface and gives a greater lowering in surface tension at the CMC. Standard free energies of micellization $G^0$mic and adsorption $G^0$ad are all negative, so micellization of the prepared inhibitors is a spontaneous process and the value increase with increasing temperatures. The standard free energy of adsorption is all negative values for inhibitors, indicating that the inhibitor BAM and ANS have the ability for adsorption at interface. On the other hand, the molecules of inhibitors favour adsorption at interface rather than sharing in the micellization processes.

**Table 1**: Physicochemical properties of the base oil

| Properties | Base oil | Test |
|---|---|---|
| Denisty (g/ml) at 15.5 C | 0.8958 | D. 1298 |
| Refactive index $nD^{20}$ | 1.4955 | D. 1218 |
| ASTM colour | 4.5 | D. 1500 |
| Kinematic viscosltycSt at 40 C at 100 C | 17.56 29.15 | D. 445 D. 455 |
| Pour point C | 15 | ASTM D. 97 |
| Molecular weight | 520 | GPC |
| Total paraffinic content, wr% | 59.353 | Urea adduction |
| Carbon residue contenty, wt% | 1.9 | ASTM D524 |
| Ash content, wt% | 0.0511 | ASTM D482 |

**Table 2**: Surface properties and free energies of surfactants at different temperatures

| Surfactants | Temperatures °C | CMC mol/l $10^3$ | $\Pi_{CMC}$ Mn/m² | $PC_{20}$ MOL/l | $x_{max} \times 10^{-11}$ mol/cm² | $A_{min}$ Nm² | $G^0$mic KJ/mol | $G^0$ad KJ/mol |
|---|---|---|---|---|---|---|---|---|
| ANS | 25 35 45 | 4 3 2 | 23 18 14 | 1.8 1.5 1 | 3.7 3.5 3.1 | 0.42 0.48 0.51 | -27.2 -27.9 -29.6 | -36.1 -36.4 -37.2 |
| BAM | 25 35 45 | 0.2 0.1 0.05 | 25 12 8 | 4 3 2.1 | 4.7 4.5 4.1 | 0.8 0.9 0.93 | -30.56 -31.2 -32.7 | -38.4 -39.9 -41.2 |

## Weight Loss Tests

The corrosion rate increases with time and tend to slightly increase after 12 days. This due to formation of stable oxide according Omar et al. [2] as shown in Figure 1. Studying the inhibition efficiencies of inhibitors at various concentrations, 0.1 mol of sulphoric acid and 25°C are shown inFigures 2-4. The efficiency was 97%, 90%, 85%, 80%, 69%, 78% and 55% for 6, 8, 10, 12, 14, 16 and 18 days respectively at 100 ppm of the inhibitor PAM. While for ANS the efficiency was 78%, 76%, 73%, 65%, 64%, 60% and 50% for 6, 8, 10, 12, 14, 16 and 18 days respectively. This means that BAM is more efficient than ANS at the same concentration and constant temperature. With further increase in inhibitor concentration, the efficiency slightly increases until 1000 ppm the efficiency steady stable. The author suggest the inhibitor tend to adsorb on the metal surface, then the inhibitor molecule tend to degradable with long times (12 days). This behavior is due to the dependence of inhibition efficiency on its surface properties. From Table 2, the area per molecule for PAM is larger than that of ANS. The effect of increasing the temperature from 25°C to 45°C for 0.1 mol of sulphuric acid at constant conditions is shown in Figure 5. For 1000 ppm of each inhibitor, the inhibition efficiency decreases with the increase in temperature from 25°C to 45°C. This can be attributed to the decrease in the protective nature of the inhibitive film formed on carbon steel surface at higher temperatures. At the same time the degree of inhibition decreases with increasing times and temperatures (Figure 5). The author thinks the hydrophilic group of each inhibitor destruct with long time and high temperature. As the results increasing temperature decrease the adsorption and consequently enhance the desorption processes.

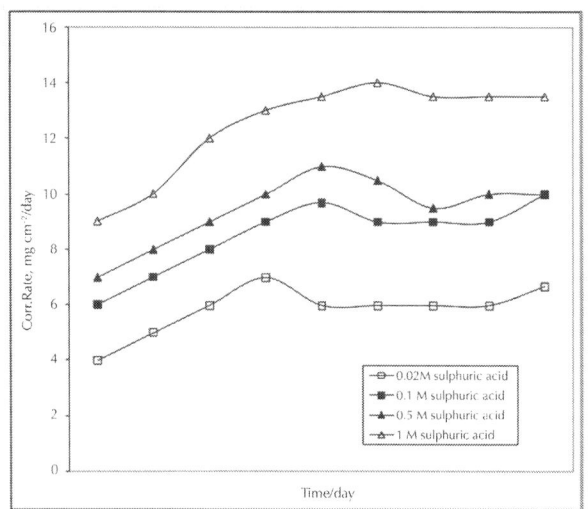

**Figure 1**: Effect of timee on corrosion rate of carbon steel in various concentration of sulphuric acid.

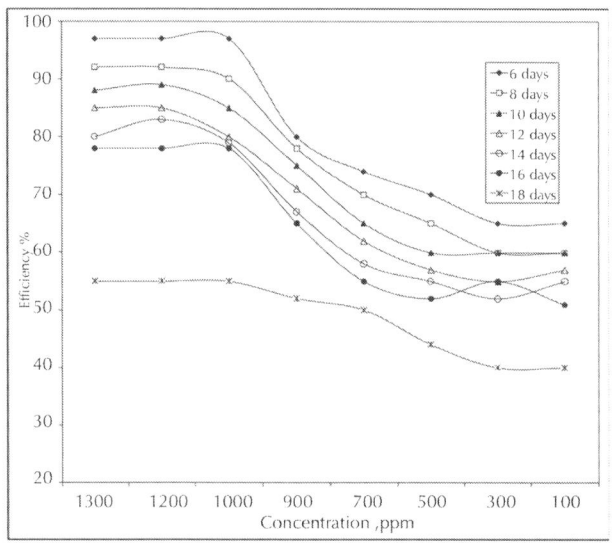

**Figure 2**: Effect of various concentration from BAM additive on its efficiency in 0.1 Mol sulphoric acid at 25 C.

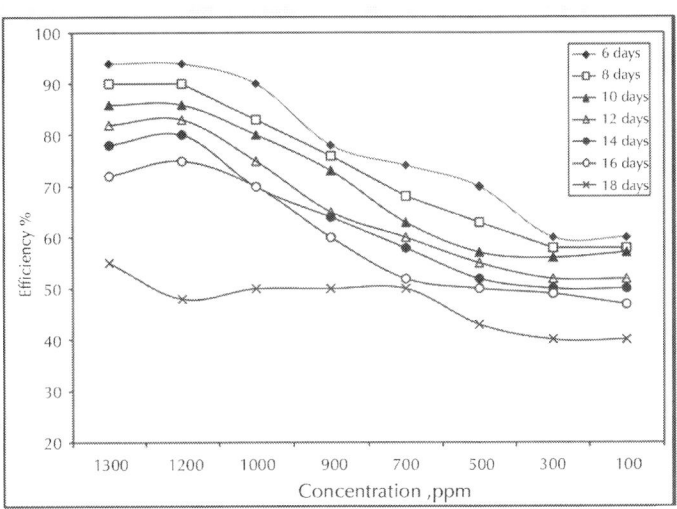

**Figure 3**: Effect of various concentration from ANS on its efficiency in 0.1 Mol sulphuric acid and 25 C.

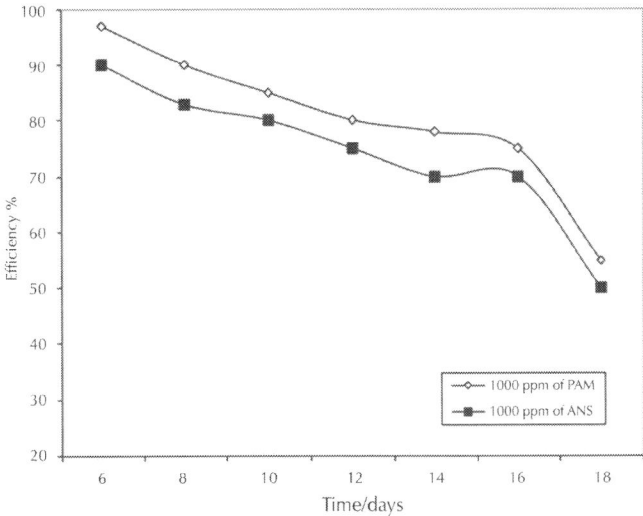

**Figure 4**: Effect of time on inhibitors efficiency at constant concentration 1000 ppm and 0.1 M of sulphuric acid.

Effect of adding inhibitors on metal working fluid at optimum conditions are shown in Figure 6. It is clear that the efficiency of each inhibitor decreases slightly with increasing time, but the degree of decreasing is slightly smaller in compare in pure acid solution as studying early. This result confirm that the synergism between inhibitors and base oil which form stable emulsion and increase degree of surface coverage on carbon steel surface.

The effect of these inhibitors on the oxidation stability of oil is given in Figure 7. The data show the additive retards the oxidation of oil for limit time and loss its efficiency after 30 hours. The additive BAM is the best; due to it has the best surface properties. The author concludes that the ability and stability of micelle is predominant factor for increase oxidation stability of oil early [5] -[8] . FromFigure 7, the total acid decrease by increasing the additive concentrations and reach the optimum value near to CMC as confirmed by the author early [5] -[8] . Further increase concentration of the additive, the oxidation stability decrease due to retardation of micelle formation, which affect on interfacial tension and degree of adsorption at interface. The azo group inhibits propagation of free radicals and terminates reaction processes of free radicals.

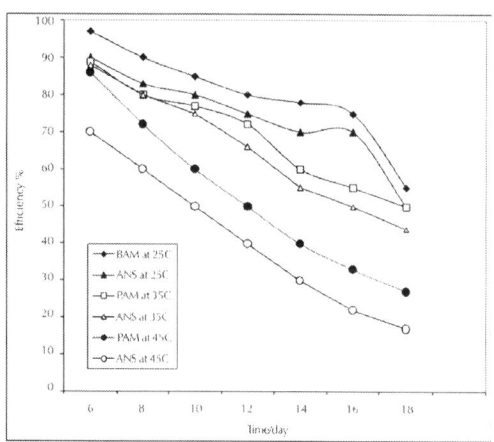

**Figure 5**: Effect of varying temperature on inhibitors efficiency in 0.1 m sulphuric acid at 1000 ppm.

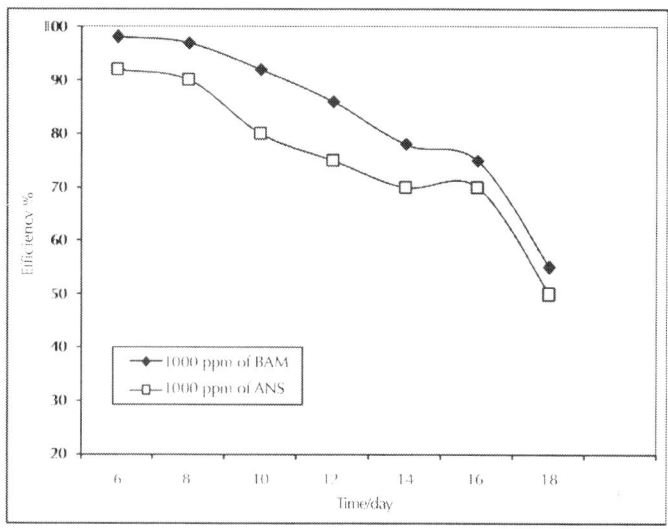

**Figure 6**: Effect of times by days on corrosion efficiency of inhibitors at concentration 1000 ppm and 25 C in metal working fluid.

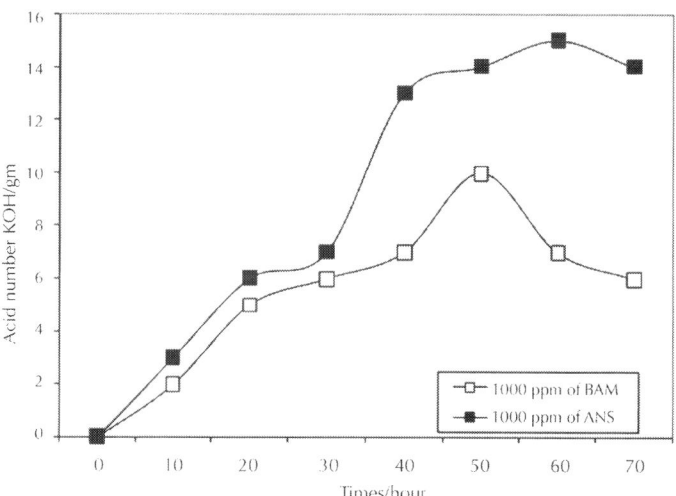

**Figure 7**: Effect of different inhibitors on oxidation stability of metal working fluid at concentration 1000 ppm and 30 C.

## CONCLUSIONS

- BAM and ANS have been shown to function as corrosion inhibitors for steel in sulphuric acid media and in metal working fluids. The chemical structure effect on degree of corrosion inhibition which reflect on surface and thermodynamic properties of inhibitor.
- The oxidation stability of oil as measured by total acid number indicates that, the oxidation inhibitor efficiency follows the order.
- BAM> ANS. These results depend on value of surface propertiese and thermodynamic parameters of inhibitors.
- The synthesized inhibitors have a multifunction for anti-corrosion and enhance oxidation stability of oil. These results depend on surface properties of the inhibitor.

## REFERENCES

1. Omar, A.M.A. (2004) Micellization and Adsorption of Anionic/Nonionic Polymeric Surfactants for Metal Working Fluid at Different. Interfaces Industrial Lubrication and Tribology, 56, 171-176.
2. Osman, M.M., Omar, A.M.A. and Al-Sabagh, A.M. (1997) Corrosion Inhibition of Benzyl Triethanol Ammonium Chloride and Its Ethoxylate on Steel in Sulphuric Acid Solution. Material Chemistry and Physics, 5013, 271-274.
3. Abdel Hamid, Z. (1997) Surface and Thermodynamic Parameters of Some Cationic Corrosion Inhibitors Anti-Corrosion Metal and Material.
4. Abdel Hamid, Z., Soror, T.Y., El Dahan, H.A. and Omar, A.M.A. (1998) New Cationic Surfactant as Corrosion Inhibitor Anti-Corrosion Metal and Material. 45, 306-311.

5. Elmehbad, N. (2013) Developments of Multifunctional Additives for High Quality Lube Oil. Journal of Power and Energy Engineering.
6. Elmehbad, N. (2013) Development Antioxidants Synthesized by Phase Transfer Catalysts for Lubricating Oil Bio Tech Conference. Expo, 12-16 May 2013.
7. Elmehbad, N. (2013) The Development and Application of Ester for Lubricating Oil by Phase Transfer Catalysts. Oral Presentation in 19th International Colloquium, Germany, 21-23 January 2013.
8. Elmehbad, N. (2014) Preparation of Anti Wear/Antioxidant Additives for High Quality Metalworking Fluid from Waste Petroleum Products. JESMAT.
9. Omar, A.M.A. and Azzam, E.M.S. (2003) Relation between Adsorption of Some Anionic Surfactants on Barite and Solution/Air Interfaces. Surface and Interface Analysis, 35, 709-713. http://dx.doi.org/10.1002/sia.1593.
10. Omar, A.M.A. and Azzam, E.M.S. (2004) Development of Electrodeposited Chromium Composite by Amphoteric Surfactant. Industrial Lubrication and Tribology, 56, 244-247.
11. Omar, A.M.A. (2001) Separation of Emulsifiable Oil from Solution by Surface Tension Control. Adsorption Science and Technology, 19, 91-100. http://dx.doi.org/10.1260/0263617011494006.

# Chapter 7

# Effect of Processing Paramters on Metal Matrix Composites: Stir Casting Process

G. G. Sozhamannan[1], S. Balasivanandha Prabu[2], and V. S. K. Venkatagalapathy[1]

[1]Department of Mechanical Engineering, Sri Manakula Vinayagar Engineering College, Pondicherry, India
[2]Department of Mechanical, Anna University, Chennai, India

## ABSTRACT

Conventional stir casting process has been employed for producing discontinuous particle reinforced metal matrix composites for decades. The major problem of this process is to obtain sufficient wetting of particle by liquid metal and to get a homogenous dispersion of the ceramic particles. In the present study, aluminium metal matrix composites were fabricated by different processing temperatures with different holding time to understand the influence of process parameters on the distribution of particle in the matrix and the resultant mechanical properties. The distribution is examined by microstructure analysis, hardness distribution and density distribution.

# INTRODUCTION

Since the early 1960, there is demand for new and improved engineering materials with advancement of modern technology interest in the areas of aerospace, automotive industries had forced a rapid development of metal matrix composites. High demands on material for better overall performance has led to extensive research and development efforts in the composites fields. Among the composites field, the aluminium based metal matrix composite materials are widely used. To meet emerging need, innovations in materials processing enabled achieving an enhancement in stiffness, realization of high strength to weight ratio, an improvement in wear resistance, maintaining strength at elevated temperatures [1,2].

Aluminium based metal matrix composites have been one of the key research areas in materials processing field in the last few decades. Most of the research work has been dealing with aluminium matrix and SiC reinforcement requiring the light weight in combination of high strength and high stiffness. This is because aluminium is lighter weight which is first requirement in most of the industries. In additionally, it provides impressive strength improvement and the thermal expansion coefficient of Al matrix composites can be adjusted by using silicon carbide, carbon and boron carbides [3]. As a result, these materials are now being rapidly utilized in industries that traditionally used metals, and these have become the forefront of research and development activity in the many related areas [4].

The foundry casting processes have been a favoured processing method as they lend themselves to the manufacture of large number of complex shaped components. Especially, the stir casting mostly used to produce the PRMMCs because it shown to be a very promising for the manufacture of near net shape composites in a simple and cost effective manner. The major problem in this technology is to obtain sufficient wetting of particle by the liquid metal and to get a homogeneous dispersion of the ceramic particles [5]. Several structural defects such as porosity, particle clustering, oxide inclusions and interfacial reactions were found to arise from the unsatisfactory casting technology [6]. Smeulders et al., (1986) [7] studied the thixotropic behaviour, development of the particle shape, and particle distribution in the metal matrix composites by using stir casting. The particles size distribution depends on the cooling rate and stirring rate. The processing temperature influence the cooling rate and stirring rate. These both parameters influence particles distribution in the matrix. At low stirring rate, diameter of cluster is appeared is smaller than higher stirring rate.

Lloyd (1989) [8] studied the processing parameters such as processing temperatures, holding time, and velocity of SiCp in liquid melts. This factors influencing the production and microstructures of the cast Al-SiCp composite, and also influence the formation reaction products ($Al_4C_3$) at the interface of Al-SiCp, which also affects the casting fluidity of composites. Zhou and Xu (1997) [6] studied the particulate reinforced metal matrix composites by gravity casting and two step methods mixing. They were found the improvement of wettability, particles distribution, and also they found SiC particles distribution in the interdendritic regions. Rohatgi et al., (1993) [9] made an attempt to study the influence of impeller geometry and baffles on the uniformity of distribution of SiC in water. This study quantitatively measures the actual SiC concentration. The objective of the study is the influence of stir casting process parameters such as processing temperature and holding time on the uniform distribution of particles and resulting mechanical properties such as Tensile, ductility, hardness and impact behaviour.

## EXPERIMENTAL WORK

The matrix material used for the current study was Al-11Si-Mg alloy, having composition average values provided by the supplier (Sargam metals Ltd., India) in weight percentage as shown in Table 1. The Al-11Si-Mg alloy has an excellent combination of mechanical properties in the cast condition. The matrix material was loaded in a graphite crucible and it was placed inside a top loaded resistance furnace at different temperature level (700°C, 750°C, 800°C, 850°C, 900°C). The SiC reinforcement (average size 40 µm, supplied by universal carborendum Ltd, India) was preheated at 1000°C for two hours before added in the matrix melt. The 10% by volume of preheated SiC powder was added in the liquid melt and the slurry was consciously stirred using a stirring. The four blade Stirrer was designed in order to produce the adequate homogenous particle distribution throughout the matrix material. Four blades stirrer setup is shown in Figure 1. The axial and radial flows are provided to avoid different stagnant zones in the liquid melt by stirrer. Stirring of the mixture is carried out at different holding time (10, 20, and 30 minutes) to achieve homogeneity of particulates. The stainless steel stirrer blade was coated with zirconia to avoid the reaction between stainless steel and Al alloys at higher temperatures. The Argon gas was supplied into the near the crucible during the stirring to avoid the formation of oxide layer on the surface of matrix melt. The Stirring speed 450 rpm was maintained throughout work. The mixture is allowed to solidify in the preheated (300°C) steel die.

**Table 1:** Composition of aluminium alloy

| Si | Cu | Mg | Ti | Ni | Li | Zn | Fe | Mn | Al |
|---|---|---|---|---|---|---|---|---|---|
| 11.12 | 0.68 | 0.73 | 0.009 | 0.002 | 0.027 | 0.002 | 0.31 | 0.32 | remains |

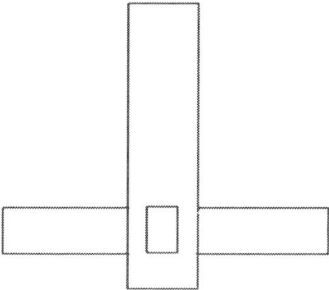

**Figure 1:** Stirrer blade setup.

# RESULTS AND DISCUSSION

## Structure of Al-10% SiCp Composites

The optical micrographs of aluminium composites reinforced with 10% of SiCp are shown in Figures 2(a)-(e). When the specimen was prepared at 700°C and 750°C, 800°C, the particles were distributed uniformly and dendrite structure was more obvious in the microsturcture (figure). There is no large pore and particles clustering existed in these SEM micrographs. However, the other two composites (at 850°C and 900°C) having pores and particles clustering. The existence of pores and particle clustering was attributed to the high viscosity and low shearing rate of the melt. The viscosity decreased and the shearing rate decreased when increasing process temperatures. Hence, the particles cluster fairly formed in the melt. This can be attributed more particles clustering in the Al matrix.

It can be seen from the figures that there is an increase the particles cluster corresponding to an increase in the processing temperatures. Which were stirred for a constant speed with longer period at higher temperature, the particles were agglomerated in the melt. Although there is an increase in the particle clustering with increase processing temperature, it was observed that the tendency for formation of particle cluster was greater in the higher holding time than in the low holding time. During the higher holding time with temperature, the geometry of the capturing of the particles does not restrict their movement inside the liquid metal as well as solidification. Also the presence of a low viscosity of liquid metal tents to physically not restricts growth of porosity. Thus, the tendency for particle cluster or porosity is high in the higher temperature with prolonged contact between matrix and reinforcement.

## Effect of Processing Temperatures on Processing Al-SiCp Composites

The effect of processing temperature is illustrated where the contact time between reinforcement and liquid Al metal with different holding time. The processing temperature is mainly influencing the change in viscosity of Al matrix and also it accelerating the chemical reactions between them. This change of viscosity was calculated theoretically by using Arrihenious equation [10].

**Figure 2:** Optical image shows particles distribution at 20 minutes holding time.

$$\eta = \eta_0 \exp\left(\frac{E}{RT}\right)$$

where $\eta_0$-viscosity of aluminium at the melting temperature, E-Activation energy for viscous flow of aluminium, R-Universal gas constant processing temperature. The viscosity of liquid Al matrix with 10% SiCp is often calculated with the Einstein function as follows [11]:

$$\eta/\eta_0 = 1 + 2.5C + 10.5C2 + \exp(A\,B)$$

in which A = 0.0023, B = 16.6, C is volume fraction of particles.

The changes in viscosity with respect to various processing temperatures are shown in Figure 3. The change of viscosity is significant, as can be seen by comparing Al alloy and Al-SiCp composites. The suspending liquid apparent viscosity of Al-SiCp is higher by nearly 38% than the Al matrix without reinforcement. But, the both viscosity is increased when processing temperature increased from 700°C to 900°C. The changes of viscosity influence the particle distribution in the Al matrix. At higher viscosity with lower temperature (below 800°C), the geometric contacts of particles is restricted by vortex of molten metal. Therefore, the particles are distributed uniformly in this range. But, at lower viscosity with higher temperature (above 800°C), the geometric of particles are captured and it was unable to restricts by the vortex of molten metal under the constant stirring speed. The conclusion can be drawn that viscosity of liquid decreased when increasing processing temperatures with increasing holding time.

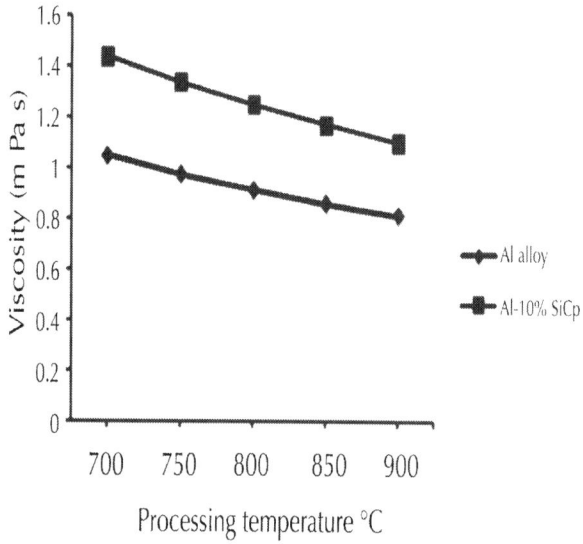

**Figure 3:** Changes of visosity as function of temeprature.

# Effect of Holding Time

Effect of the holding time helps in the Al-SiCp composites mainly two ways: to distribute the particles in the liquid, and to create perfect interface bond between reinforcement and matrix. The holding time between matrix and reinforcement is considered as important factor in the processing of composites. When the holding time is 10 minutes, the particles are distributed uniformly in the matrix at 700°C, 750°C and 800°C. The liquid matrix has sufficient viscosity in the temperature range, and velocity of particles flow is small. The similar results are observed in the 20 minutes holding time. In the case of 30 minutes holding time, the liquid has sufficient viscosity at lower temperature (<800°C) but the contacts time between matrix and reinforcement too large. During this period, the particles are distributed uniformly in liquid even though some of particles form cluster which could be also are located in the matrix region. A vortex created during the stirring can suck the air or gas bubbles in to the liquid metal. As the results, the particles were attached with air bubbles to form the particles cluster in the matrix. At higher temperature (>800°C), more particles cluster are found in the composite bar.

# EFFECT OF PROCESSING PARAMETERS ON MECHANICAL PROPERTIES OF AL-SICP

## Tensile Properties

Figure 4 shows the variation in the ultimate tensile strength as a function of temperature is illustrated for Al-10% SiCp composites with different holding time.

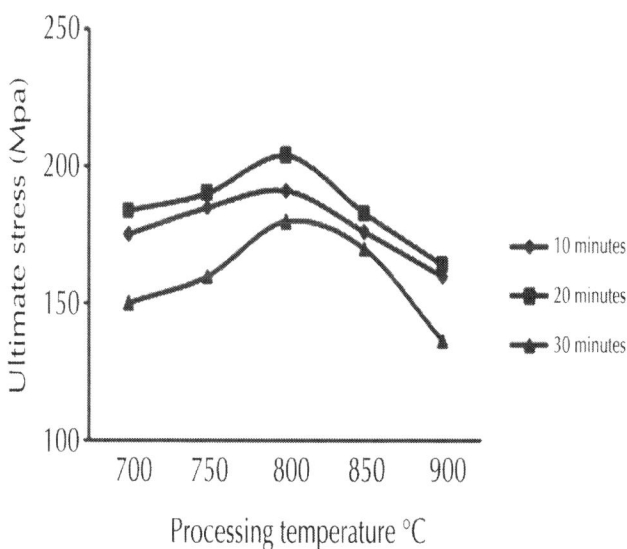

**Figure 4:** Effect of processing parameters on tensile strength of Al/SiCp.

The ultimate strength of Al-10% SiCp composite has been increased and reached a maximum as the processing temperature changes from 700°C to 800°C, then began decrease with further increase of processing temperature from 800°C to 900°C. These composites exhibited different tensile behaviors. The overall strength of composites is influenced by distribution of particles in the matrix.

It is show that the SiC particles are uniformly distributed in the matrix when the processing temperatures are at 700°C to 800°C for 10 minutes holding time. It is obvious that the reduction in viscosity on increasing the processing temperatures which is shown in figure during the tensile load, the presences of particles hinder the dislocation movement during plastic deformation and it exhibits the isotropic properties in the matrix. Therefore, the ultimate strength is increased gradually. Further increase in temperature from 800°C to 900°C, the particles cluster associated with porosity is formed in the liquid matrix due to the vortex of molten stirrer which entrapped the gases inside the matrix. The viscosity also decreased further. The clustering of particles has considerable effect on strength and plastic behavior of composites. Elastic response of the composite is not much affected due to clustering nature

## Hardness Distribution

The Brinell hardness number was measured along length of the cast specimen at an interval of 1 cm. The low temperature with holding time hardness values at some places is minimum it close to harness values of Aluminium and some places is more. The high values is obtained from the places where the particles is accumulation more and lower hardness values is obtained from places where SiC particles where absent. The Figure 5 shows the hardness number distributions along length of the cast specimens. The SiC particles added to the aluminium alloy matrix have a satisfactory effect in improving the hardness of the composites. This is to be expected since aluminium is a soft material and the SiC particles being hard, contribute positively to the hardness of the composite. The presence of stiffer and stronger particles leads to constrain the plastic deformation of the matrix during the hardness test. But, constrains to the plastic deformation depend on the distribution of the particles in the matrix. Some local regions in the composite show remarkable changes, if the particle accumulated in a particular place the values are higher, and if the particles are absent in some places the values are lower. The hardness measurements were taken in cast specimen to understand the distribution of the SiC particle in the Al alloy at different processing conditions.

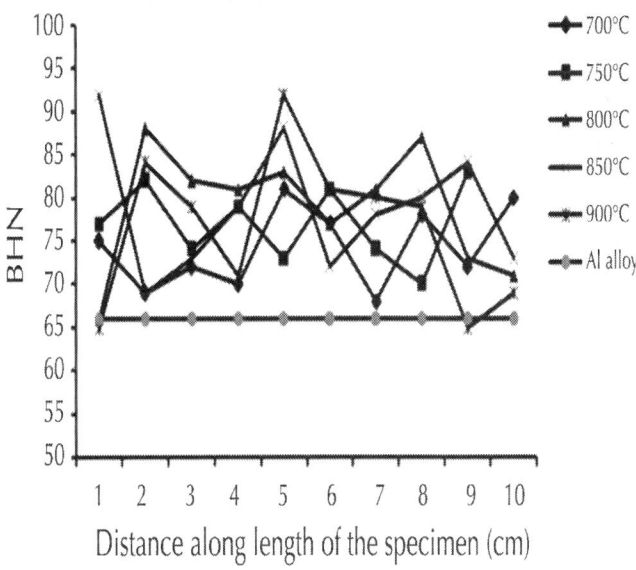

**Figure 5:** Effect processing temperatures on Hardness of Al/SiCp.

## Impact Test

The results of the charpy impact tests for Al-10% SiCp composites fabricated with different processing temperatures with holding time are given in Table 2" target="_self"> Table 2. The test results revealed that the impact energy of Al-SiCp is mainly depends on the distribution of the particles in the matrix. It is interesting to note there is a little variation in the value for the different processing conditions as shown in Table 2. The impact values are slightly increases with increasing the processing temperatures. The decrease of impact values at lower temperature in the Al-SiCp composites can be attributed to the presence of brittle SiCp which may act as stress concentration areas, where the particles are distributed uniformly.

**Table 2:** Impact value of Al-SiCp composites

| Processing temperatures °C | Impact values with different holding time (J) | | |
|---|---|---|---|
| | 10 mins. | 20 mins. | 30 mins. |
| 700 | 3 | 4 | 6 |
| 750 | 3 | 6 | 6 |
| 800 | 4 | 6 | 8 |
| 850 | 5 | 8 | 8 |
| 900 | 5 | 10 | 10 |

## CONCLUSIONS

The following conclusion can be reached based on the present investigation:

- From the microstructure analysis, the particles were distributed uniformly in the processing temperature 750°C and 800°C. The particles agglomerations were found in the processing temperature of 700°C, 850°C = and 900°C due to the changes of viscosity in liquid Al matrix.

- The viscosity of Al matrix decreases with increased processing temperatures. The suspending liquid viscosity of Al-SiCp is higher by nearly 38% than the Al matrix without reinforcement.
- The tension test revealed that ultimate strength increased gradually up to 800°C and starts to decrease gradually due to the distribution in the Al matrix.
- The Ultimate strength of metal matrix composite decreases with increasing holding time. It is revealed that holding time influences the viscosity of liquid metal, particles distribution and also induces some chemical reaction between matrix and reinforcement.
- The hardness values increases more or less linearly with increasing of processing temperatures from 750°C to 800°C at 20 minutes holding time.

# REFERENCES

1. T. P. Rajan, R. M. Pillai and B. C. Pai, "Review Reinforcement Coatings and Interfaces in Aluminium Metal Matrix Composites," Journal Material Science, Vol. 33, No. 14, 1998, pp. 3491-3503. doi:10.1023/A:1004674822751
2. M. Jayamathi, S. Seshan, S. V. Kailas, K. Kumar and T. S. Srivatsan, "Influence of Reinforcement on Microstructure and Mechanical Response of a Magnesium Alloy," Current Science, Vol. 87, No. 9, 2004, pp. 1218-1231.
3. V. K. Lindroos and M. J. Talvitie, "Recent Advances in Metal Matrix Composites," Journal of Material Processing Technology, Vol. 53, 1995, pp. 273-284.
4. L. M. Tham, M. Gupta and L. Cheng, "Effect of Limited Matrix-Reinforcement Interfacial Reaction on Enhancing the Mechanical Properties of Aluminium-Silicon Carbide Composites," Acta Materialia, Vol. 49, No. 16, 2001, pp. 3243-3253. doi:10.1016/S1359-6454(01)00221-X

5. J. Hashim, L. Looney and M. S. J. Hashmi, "Metal Matrix Composites: Production by the Stir Casting Method," Journal of Materials Processing Technology, Vol. 119, No. 1-3, 1999, pp. 329-335. doi:10.1016/S0924-0136(01)00919-0
6. W. Zhou and Z. M. Xu, "Casting of SiC Reinforced Metal Matrix Composites," Journal of Materials Processing Technology, Vol. 63, No. 1-3, 1997, pp. 358-363. doi:10.1016/S0924-0136(96)02647-7
7. R. J. Smeulders and F. H. Mischgofsky, "Direct Microscopy of Alloy Nucleation, Solidification and Ageing (Coarsening) during Stir Casting," Journal of Crystal Growth, Vol. 76, No. 1, 1986, pp. 151-169. doi:10.1016/0022-0248(86)90021-7
8. D. J. Lloyd, "The Solidification Microstructure of Particulate Reinforced Aluminium/SiC Composites," Composite Science and Technology, Vol. 35, No. 2, 1989, pp. 159-179.
9. P. K. Rohatgi, S. Ray, R. A. Sthena and C. S. Narendranath, "Interface in Cast Metal Matrix Composites," Materials Science and Engineering, Vol. 162, No. 1-2, 1993, pp. 163-174. doi:10.1016/0921-5093(90)90041-Z
10. F. Miani and P. Matteazzi, "Estimation of Viscosity in under Cooled Liquid Metal Alloys," Journal of Noncrystalline Solids, Vol. 143, 1992, pp. 140-146. doi:10.1016/S0022-3093(05)80561-7
11. J. Wang, Q. X. Guo, M. Nishio, H. Ohawa, D. Shu and K. Li, "The Apparent Viscosity of Fine Particle Reinforced Composite Melt," Journal of Materials Processing Technology, Vol. 136, No. 1, 2003, pp. 60-63. doi:10.1016/S0924-0136(02)00919-6

# Chapter 8

# The Effect of Drilling Fluids and Crude Oil on Some Chemical Characteristics of Soil and Crops

Ivica Kisic[a], Sanja Mesic[b], Ferdo Basic[a], Vladislav Brkic[b], Milan Mesic[a], Goran Durn[c], Zeljka Zgorelec[a], Lidija Bertovic[b]

[a]University of Zagreb, Faculty of Agriculture, Svetosimunska cesta 25, 10 000 Zagreb, Croatia
[b]INA-NAFTAPLIN, Subiceva 29, 10 000 Zagreb, Croatia
[c]University of Zagreb, Faculty of Mining, Geology and Petroleum Engineering, Pierottijeva 6, 10 000 Zagreb, Croatia

## ABSTRACT

A four-year pot trial was set up to determine, as precisely as possible, the influence of increased levels of total petroleum hydrocarbons

(TPH) upon soil and plants grown. In eight treatments, clean soil and different doses of drilling fluids and crude oil were applied. The changes in some chemical parameters of soil, plant density and crop yields were investigated. The influence of the studied indicators on the achieved plant density and crop yield was strongest in the first trial year. Drilling fluids had a stronger impact on the chemical properties of the studied soil, while plant density and yield were more strongly affected by crude oil. Upon application of drilling fluids and crude oil, the soil pH, contents of organic matter (OM) and heavy metals (HM) varied very little throughout the trial period, whereas the soil levels of total petroleum hydrocarbons, mineral oils (MO) and polycyclic aromatic hydrocarbons (PAHs) were significantly reduced after the first trial year.

# INTRODUCTION

TPH introduction into the soil environment can occur from pipeline blow-outs, waste deposition after drilling oil and gas wells, road accidents, leaking underground storage tanks, land farming and uncontrolled landfill (Chaineau et al., 2003). With hindsight, it is interesting to note that it was once thought that a certain amount of TPH could serve as fertilizer and stimulate plant and animal growth (Carr, 1919, Murphy and Riley, 1929 and Mackin, 1950). The reasons for reduced plant growth in soils contaminated by TPH range from direct toxic effects of oil on plants (Baker, 1970 and Kyung-Hwa et al., 2004), lack of germination due to the lack of viable seeds (Ekundayo et al., 2001 and Ogboghodo et al., 2004a), reduced germination (Dorn and Salanitro, 2000), and unsatisfactory soil conditions. Soil conditions may be poor due to insufficient aeration caused by decreased air-filled pore space, and increased oxygen demand caused by oil-decomposing microorganisms, as well as a reduction in the level of available plant nutrients) (De Jong, 1980). According to available literature, the effects of TPH on soil used to be studied in two different ways. One investigation involved the addition of certain amounts of TPH into clean soil after sowing (Sarkar et al., 2005, Okolo et al., 2005 and Agbogidi et al., 2007). In the second investigation, the studied crop was sown into previously TPH contaminated soil (De Jong, 1980, Akaninwor et al., 2007 and Shahriari et al., 2007). This paper presents the results of experiment made by mixing the clean soil (treatment I) with a certain percentage of drilling fluids (treatments: II, III, IV and V) and then sowing the crops studied. The Experiment also included treatments in which clean soil was mixed with crude oil (treatments VII and VIII) taken from oil wells in the immediate vicinity of a pipeline breakage. Treatment VI represents soil that was hauled as "replacement" soil to the site of the pipeline breakage.

## MATERIAL AND METHODS

A pot trial was set up in the greenhouse of the Faculty of Agriculture in Zagreb, Croatia, in the autumn 2003. The trial included 4 replications of each following treatments:

- Control (clean soil) taken in the immediate vicinity of a pipeline breakage site
- Drilling fluids — taken from the central waste pit of the oil/gas field
- 1/2 clean soil + 1/2 drilling fluids (6 kg clean soil + 6 kg drilling fluids)
- 2/3 clean soil + 1/3 drilling fluids (8 kg clean soil + 4 kg drilling fluids)
- 3/4 clean soil + 1/4 drilling fluids (9 kg clean soil + 3 kg drilling fluids)
- Soil hauled to the pipeline breakage site
- 2/3 clean soil + 1/3 crude oil (8 kg clean soil + 4 l crude oil taken from the oil well)
- 3/4 clean soil + 1/4 crude oil (9 kg clean soil + 3 l crude oil taken from the oil well)

The Experiment ran from 2003 to 2007. After the trial was prepared according to the above methodology, the crop sowing began. According to the crop sequence, winter wheat (*Triticum aestivum* L.) was sown on 14 October 2003 and 26 October 2005; winter barley (*Hordeum vulgare* L.) was sown on 21 October 2004 and 27 October 2006; soybean (*Glycine hyspida* L.) was sown on 29 June 2005 and 3 July 2006. The greenhouse trial was set up in pots with 4 replications; the trial pot area was 0.15 m$^2$. Standard agro-technical practices were applied to the pots — chemical protection and fertilization of crops. Basic soil chemical analyses (soil pH and organic matter content), TPH and MO contents, were performed once a year. HM and PAHs were determined in 2003, 2005, and 2007. Soil samples (for above mentioned analyses) were taken after crop harvest and prior to sowing the next crop.

The research objective was to investigate the possibility of field crops production on TPH contaminated soils, and to identify the effect of such contamination on plant density and crop yield by determining:

- changes in soil chemical complex (soil pH, OM, TPH and MO, HM and PAHs),
- effects of different TPH and PAHs levels on emergence, plant density and yield of crops grown.

The investigation results were statistically processed by ANOVA and the t-test to estimate the significance of the differences between the treatments and control. The methods used to determine the studied parameters are given in Table 1.

**Table 1:** Methods used in investigations

| Analysis | Method |
| --- | --- |
| Soil sampling | ISO 10381-1-8 (2001–2006) |
| Preparation of soil samples for physical and chemical analyses | ISO 11464:2004 |
| Pretreatment of samples for determination of organic contaminants | ISO 14507:2003 |
| Preparation of laboratory samples from large samples | ISO/DIS 23909:2007 |

| | |
|---|---|
| Determination of organic (TOCIOM) and total carbon (TC) by dry digestion (elemental analysis) | ISO 10694:2004 |
| Determination of total nitrogen by dry digestion (elemental analysis) | ISO 13878:2004 |
| Determination of pH values (KCl) 1:2.5 | ISO 10390:2004 |
| Extraction of aqua regia soluble elements | ISO 11466:2004 |
| Determination of Zn, Pb, Cd, Co, Ni, Cr and Cu using AAS | ISO 11047:2004 ISO 11885:1998 |
| Determination of As, Ba, Mo, V and Hg using ICPMS | ISO/DIS 22036:2006 |
| Determination of total petroleum hydrocarbons and mineral oils in soil --gas chmmatography | ISO 16703:2004 |
| Determination of polycyclic aromatic hydrocarbons | ISO 18287:2005 EPA 550 |

# RESULTS AND DISCUSSION

The major physical and chemical characteristics of water-based drilling fluids (muds) and crude oil applied in the trial are given in Table 2 and Table 3.

**Table 2:** Content of heavy metals in drilling fluids applied in the trial[a]

| | Cd, mg/kg | Hg, mg/kg | Pb, mg/kg | As, mg/kg | Ni, mg/kg | Cu, mg/kg | Cr, mg/kg | Zn, mg/kg | Ha, mg/kg | Ca, g/kg |
|---|---|---|---|---|---|---|---|---|---|---|
| Min | 8.8 | 2.6 | 187 | 35.8 | 27.5 | 26.8 | 47.2 | 139 | 1988 | 6.85 |
| Max | 11.0 | 4.8 | 358 | 49.0 | 39.5 | 41.6 | 68.2 | 295 | 2674 | 11 |
| Average | 9.6 | 3.8 | 219 | 41.2 | 34.3 | 31.2 | 57.8 | 206 | 2373 | 9.03 |

Survey of 20 drilling fluids samples taken from the central waste pit in the 011 field where pipeline breakage occurred.

**Table 3:** Some characteristics of crude oil applied in the trial[a]

|  | Sum, vol. % | Density at 15 °C, g/cm³ | Viscosity at 37.8 °C mm²/s |
|---|---|---|---|
| Light gasoline | 3.87 | 0.6702 | |
| Light gasoline + heavy gasoline | 19.27 | 0.7447 | |
| Kerosene | 9.53 | 0.8151 | |
| Gas oil | 14.28 | 0.8446 | |
| High viscosity lubricant oil | 9.94 | 0.8548–0.8705 | 7.5–20.6 |
| Medium viscosity lubricant oil | 6.88 | 0.8705–0.8832 | 20.6–43 |
| High viscosity lubricant oil | | | Over 43 |
| Residue | 39.17 | 0.9530 | |
| Loss | 0.93 | | |

| Paraffins | Naphthenes | Aromatics | Olefins | Sum |
|---|---|---|---|---|
| 53.34 | 38.73 | 7.93 | | 100.00 |

| Metals, mg/kg | | | | | | | | |
|---|---|---|---|---|---|---|---|---|
| Na– | Ca– | Mg– | Ni– | Pb– | Fe– | Zn– | Cr– | v– |
| 0 | 12.94 | 0.44 | 6.31 | 0 | 1.04 | 0.49 | 0.24 | 0 |

Average value of crude oil taken from the well near pipeline breakage.

## Changes in Soil PH, Content Of Organic Matter, Total Carbon, Total Nitrogen and C/N Ratio

Trial results show heterogeneity in soil pH and organic matter content. Table 4 and Fig. 1 present statistically significant differences in soil pH in all treatments, compared to control treatment (clean soil — treatment I). An increase in pH was recorded in treatments involving application of drilling fluids (100% or some other percent), which was expected because drilling fluids are rich in calcium (Table 2). Increased calcium levels in drilling fluids are a direct result of the use of calcium as an additive for preventing corrosion of oil/gas pumping pipes and for raising fluid density during drilling (Carls et al., 1995 and Bauder et al., 2005). $CaCO_3$ is also commonly added to bind fluids during remediation of pipeline breakages or other incidents. In treatments where crude oil was applied, pH changed statistically significantly compared to control treatment. At the pipeline breakage site (treatment VI), where clean soil was brought, pH also increased as a consequence of the different pH value of the hauled soil. Similar and statistically significant differences were recorded for OM (Table 4 and Fig. 2). The highest organic matter content was recorded in treatment II (100% drilling fluids). In other treatments, in accordance with the levels of drilling fluids added, statistically significantly higher OM values were recorded compared to the control treatment. These findings are attributed to the chemical composition of drilling fluids and crude oil. The analytical method of organic matter determination is based on the Dumas method of dry digestion, in which total carbon is determined in the sample, and then total organic carbon is determined upon treatment with dilute HCl solution. Multiplying TOC (Total Organic Carbon) by factor 1.85 gives OM; in other words, the OM percentage is obtained by the analysis of TOC, which naturally contains TPH. For this reason, the statistically significantly lower content of TOC in control treatment compared to all other treatments is not surprising (Table 4 and Fig. 3).

**Table 4:** Changes in soil pH, organic matter, total carbon, total nitrogen and C/N ratio

| Treatment year | I | II | III | IV | V | VI | VII | VIII |
|---|---|---|---|---|---|---|---|---|
| Soil pH | | | | | | | | |
| 2003 | 6.51 | 6.67** | 6.82** | 6.78** | 6.43** | 7.13** | 6.51 | 6.42** |
| 2005 | 6.47 | 6.93** | 6.89** | 6.89** | 6.64** | 7.07** | 6.54** | 6.44** |
| 2007 | 6.17 | 7.08** | 7.03** | 7.12** | 6.75** | 7.38** | 6.71** | 6.21** |
| Organic matter, % | | | | | | | | |
| 2003 | 2.13 | 5.57** | 4.18** | 3.24** | 3.64** | 3.54** | 1.92** | 2.23* |
| 2005 | 2.43 | 5.39** | 4.05** | 3.44** | 3.54** | 3.44** | 1.82** | 2.33* |
| 2007 | 2.33 | 5.31** | 3.74** | 3.54** | 3.44** | 3.24** | 2.05** | 3.44** |
| Carbon, % | | | | | | | | |
| 2003 | 1.74 | 4.90** | 2.94** | 2.62** | 2.04** | 1.96** | 1.90** | 1.83** |
| 2005 | 1.67 | 4.85** | 2.71** | 2.56** | 2.12** | 1.81** | 1.85** | 1.91** |
| 2007 | 1.69 | 4.89** | 2.64** | 2.57** | 2.16** | 1.88** | 1.91** | 1.80** |
| Nitrogen, % | | | | | | | | |
| 2003 | 0.22 | 0.18 | 0.17 | 0.16 | 0.17 | 0.12 | 0.14 | 0.15 |
| 2005 | 0.24 | 0.21 | 0.18 | 0.18 | 0.19 | 0.13 | 0.15 | 0.16 |
| 2007 | 0.21 | 0.19 | 0.19 | 0.18 | 0.20 | 0.15 | 0.15 | 0.17 |
| Carbon/nitrogen ratio | | | | | | | | |
| 2003 | 8 | 27** | 17** | 16** | 12** | 16** | 14** | 12** |
| 2005 | 7 | 23** | 15** | 14** | 11** | 14** | 12** | 12** |
| 2007 | 8 | 26** | 14** | 14** | 11** | 13** | 13** | 11** |

*,**Significant at the 0.05 and 0.01 levels of probability respectively.

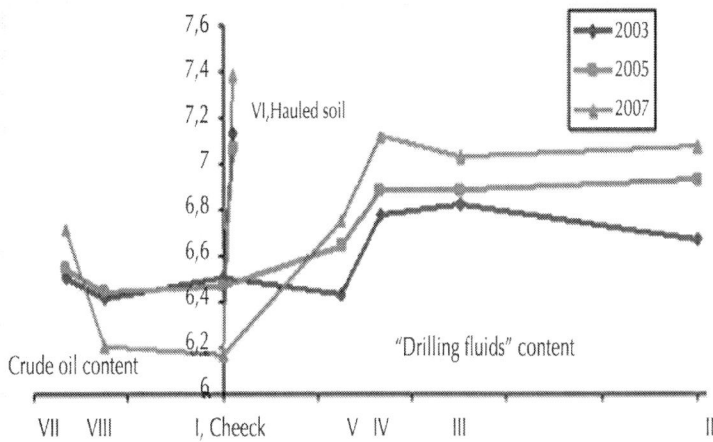

**Figure 1:** Effects of drilling fluids and crude oil on soil pH.

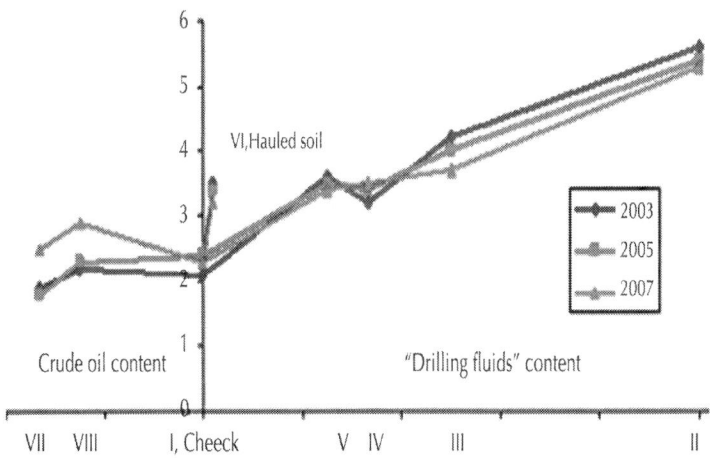

**Figure 2:** Effects of drilling fluids and crude oil on content of organic matter in soil (OM), %.

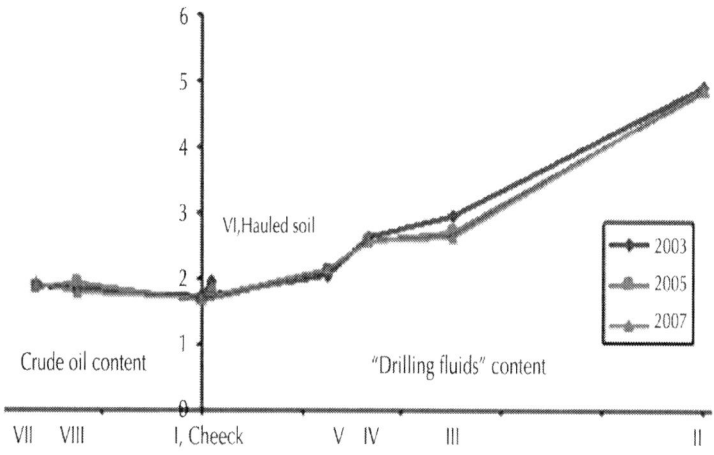

**Figure 3:** Effects of drilling fluids and crude oil on total carbon content in soil (TC), %.

Due to elevated carbon content and lower nitrogen content (Table 4) in treatments where drilling fluids were applied (treatments II and III), significant changes also occurred in the C/N ratio. The decrease in total nitrogen (Table 4 and Fig. 4) with an increase in TPH, MO (Table 5) and PAHs (Table 6) may be due to temporal immobilization of this nutrient by microbes, which might have increased in population. Jobson et al. (1974) and De Jong (1980) report on unfavourable carbon to nitrogen ratios in their investigations. Nitrogen addition with mineral fertilizers (Kirkpatrick et al., 2006) or organic soil improvers (Callaham et al., 2002, Ogboghodo et al., 2004b and Adedokun and Ataga, 2007) may enhance the development of microbiological processes in soil and thereby improve the C/N ratio (Fig. 5).

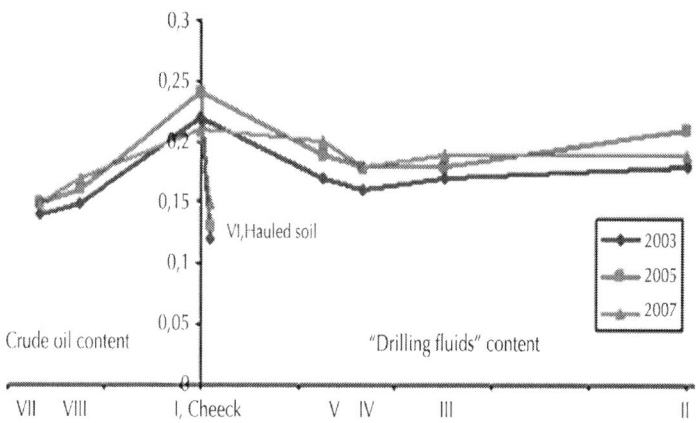

**Figure 4:** Effects of drilling fluids and crude oil on total nitrogen content in soil (TN), %.

**Table 5:** Changes in total petroleum hydrocarbon (TPH) and mineral oils (MO)

| Treatment 1 year | I | II | III | IV | V | VI | VII | VIII |
|---|---|---|---|---|---|---|---|---|
| Total petroleum hydrocarbons-TPH, g kg$^{-1}$ | | | | | | | | |
| 2003 | 0.44 | 76.1** | 24.5** | 14.7** | 6.3** | 1.1** | 2.4** | 3.1** |
| 2005 | 0.06 | 8.0* | 4.1** | 2.9** | 2.2** | 3.8** | 1.8** | 2.5** |
| 2007 | 0.08 | 5.9** | 3.5** | 2.1** | 1.6** | 2.7** | 1.3** | 2.2** |
| Mineral oils — MO, g kg$^{-1}$ | | | | | | | | |
| 2003 | 0.23 | 49.6** | 7.4** | 4.2** | 3.5** | 0.4** | 0.6** | 0.4** |
| 2005 | 0.01 | 1.6** | 1.1** | 0.9** | 0.2** | 1.5** | 0.2** | 0.8** |
| 2007 | 0.03 | 1.7** | 1.1** | 0.7** | 1.6** | 1.2** | 0.8** | 0.9** |

*,**Significant at the 0.05 and 0.01 levels of probability respectively.

**Table 6:** Soil contamination by polycyclic aromatic hydrocarbons (PAHs)

| Type of PAHs | Four-ring, mg kg⁻¹ dry soil | | | | | Five- or six-ring. mg kg⁻¹ dry soil | | | Total |
|---|---|---|---|---|---|---|---|---|---|
| | Flour-an-thene | Pyrene | Benzo (b) fluoran-thene | Benzo (k) fluor-anthene | Benzo (a) py-erene | Dibenzo (ah) an-thracene | Benzo (ghi) per-ylene | Indeno (1,2,3-cd) pyrene | |
| Treatment | | | | | | | | | |
| **2003** | | | | | | | | | |
| I | < 0.01 | < 0.01 | < 0.01 | < 0.01 | < 0.01 | < 0.01 | < 0.01 | < 0.01 | < 0.01 |
| II | 5 | < 0.01 | 42 | < 0.01 | < 0.01 | 1.5 | < 0.01 | < 0.01 | 48.5** |
| III | 6 | 8 | < 0.01 | < 0.01 | < 0.01 | < 0.01 | < 0.01 | < 0.01 | 14.00** |
| IV | < 0.01 | < 0.01 | < 0.01 | < 0.01 | 2.5 | 3.2 | 4.1 | 0.5 | 10.30** |
| V | 1 | 1.1 | < 0.01 | < 0.01 | 2.1 | 1.5 | 1.4 | 1.4 | 8.50** |
| VI | 0.1 | 0.1 | 0.1 | 0.1 | 0.9 | 0.1 | 014 | 0.1 | 1.64** |
| VII | 8.1 | 5.9 | 2.4 | 4.7 | 9.1 | 4.7 | 5.7 | 3.1 | 43.70** |
| VIII | 6.7 | 4.1 | 2.7 | 3.8 | 10.1 | 5.4 | 6.1 | 2.8 | 41.70** |
| **2005** | | | | | | | | | |
| I | < 0.01 | < 0.01 | < 0.01 | < 0.01 | < 0.01 | < 0.01 | < 0.01 | < 0.01 | < 0.01 |
| II | 0.3 | < 0.01 | 4.1 | < 0.01 | < 0.01 | 0.2 | < 0.01 | < 0.01 | 4.6** |
| III | < 0.01 | < 0.01 | 0.03 | < 0.01 | < 0.01 | < 0.01 | < 0.01 | < 0.01 | 0.03** |
| IV | < 0.01 | < 0.01 | < 0.01 | < 0.01 | < 0.01 | < 0.01 | < 0.01 | < 0.01 | < 0.01 |
| V | < 0.01 | < 0.01 | 0.03 | 0.02 | < 0.01 | < 0.01 | < 0.01 | < 0.01 | 0.05** |
| VI | < 0.01 | < 0.01 | 0.01 | 0.01 | < 0.01 | < 0.01 | < 0.01 | < 0.01 | 0.02* |
| VII | 0.4 | < 0.01 | 0.2 | 0.5 | 0.8 | < 0.01 | 0.1 | 0.1 | 2.1** |
| VIII | 0.4 | < 0.01 | 0.9 | 0.1 | 1.7 | 0.1 | 0.1 | < 0.01 | 3.3** |
| **2007** | | | | | | | | | |
| I | < 0.01 | < 0.01 | < 0.01 | < 0.01 | < 0.01 | < 0.01 | < 0.01 | < 0.01 | < 0.01 |
| II | < 0.01 | < 0.01 | < 0.01 | < 0.01 | < 0.01 | < 0.01 | < 0.01 | < 0.01 | < 0.01 |
| III | < 0.01 | < 0.01 | < 0.01 | < 0.01 | < 0.01 | < 0.01 | < 0.01 | < 0.01 | < 0.01 |
| IV | 0.01 | < 0.01 | 0.03 | 0.02 | < 0.01 | < 0.01 | < 0.01 | < 0.01 | 0.05** |
| V | < 0.01 | < 0.01 | < 0.01 | < 0.01 | < 0.01 | < 0.01 | < 0.01 | < 0.01 | < 0.01 |
| VI | < 0.01 | < 0.01 | < 0.01 | < 0.01 | < 0.01 | < 0.01 | < 0.01 | < 0.01 | < 0.01 |
| VII | < 0.01 | < 0.01 | < 0.01 | < 0.01 | < 0.01 | < 0.01 | < 0.01 | < 0.01 | 0.02* |
| VIII | < 0.01 | < 0.01 | 0.02 | 0.02 | < 0.01 | < 0.01 | < 0.01 | < 0.01 | 0.04** |

*,**Significant at the 0.05 and 0.01 levels of probability respectively.

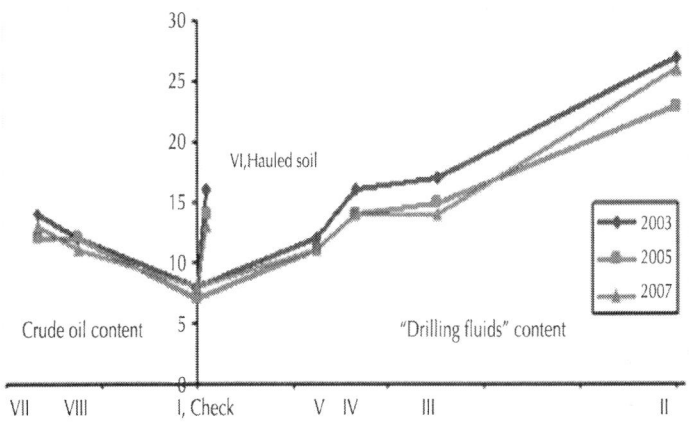

**Figure 5:** Effects of drilling fluids and crude oil on soil C/N ratio.

# Total Petroleum Hydrocarbons and Mineral Oils

Crude oil and petroleum products are complex mixtures of hundreds of hydrocarbon compounds, ranging from light, volatile, short-chained organic compounds to heavy, long-chained, branched compounds. Soil contamination by TPH and MO is a growing concern since it may be a source of groundwater contamination (Healy et al., 2001 and Asia et al., 2007). Contaminated soils can reduce the usability of land (Carls et al., 1995), and weathered petroleum residuals may remain bound to soils for years (Sarkar et al., 2005). MO contains hundreds of hydrocarbon compounds, including a substantial fraction of nitrogen and sulphur-containing compounds. Hydrocarbons are mainly mixtures of straight and branched chain hydrocarbons (alkanes), cycloalkanes and aromatic hydrocarbons.

Soil contamination by TPH and MO is illustrated in Table 5 and Fig. 6 and Fig. 7. Compared to the control, statistically higher levels of TPH and MO were found in all treatments during the trial period. This was most significant in the first trial year. In the second, third, and fourth years, the TPH and MO contents decreased in relation to the first year, but they were still (statistically) significantly higher compared to the control. Residual TPH and MO degraded very slowly after the first year and their levels remained constant in the remaining trial period. These findings indicate that in the first trial year, most of the TPH and MO, with the volatile aromatic fraction prevalent in their composition (Table 3), evaporate into the air or degrade through microbiological processes (Jobson et al., 1974, Yong et al., 1992, Chaineau et al., 2003 and Ogboghodo et al., 2004b). Depending on the soil's mechanical composition, TPH and MO bind strongly to the soil adsorption complex, and only slightly and slowly percolate towards deeper horizons (Rhykerd et al., 1999 and Pezeshki et al., 2000). TPH and MO left over in the soil in amounts below 5 g/kg of soil, degrade very slowly and, according to the available research results, are not harmful to crops. More precisely, the studied crops grow without much greater restraint on soil containing less TPH and MO than 5 g/kg soil. Akaninwor et al. (2007) and Chaineau et al. (2003) came to similar conclusions in their investigations. Kyung-Hwa et al. (2004) report that 10 g/kg soil was toxic to the studied crops, but also that no phytotoxicity was determined in soil containing up to 1 g/kg hydrocarbons. Our previous investigations (Kisic et al., 2005) have shown that the rate of TPH and MO degradation in soil was mainly connected with the kind, quantity, and type of crude oil or drilling fluid, OM content and soil mechanical composition (clayey or sandy soil), soil moisture, together with the season in which increased TPH content was spilled into the soil.

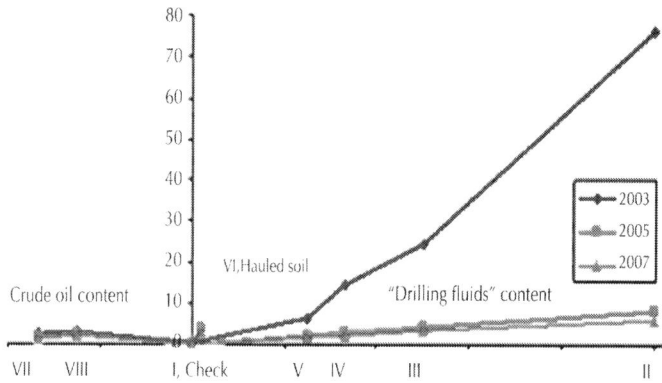

**Figure 6:** Effects of drilling fluids and crude oil on total petroleum hydrocarbons (TPH) in soil, g/kg.

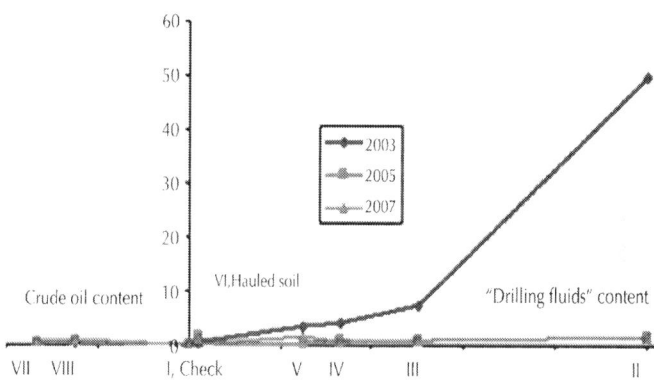

**Figure 7:** Effects of drilling fluids and crude oil on mineral oil (MO) in soil, g/kg.

## Polycyclic Aromatic Hydrocarbons

Crude oil and drilling fluids are a complex mixture of hydrocarbons containing PAHs and non-hydrocarbon(s) compounds including HM, which are potentially phytotoxic (Kelsey and Alexander, 1997 and Samanta et al., 2002), and may interfere with normal plant development and reproduction (Mendelssohn et al., 1990 and Adam and Duncan, 2002). PAHs are large group of polycyclic hydrocarbons containing one or more benzene rings. According to the World Health Organization's (WHO) recommendations for consideration and interpretation, measurements of emission into the atmosphere should involve only benzo(a)pyrene, as the best known and most studied among several thousands of PAHs, whose toxic, carcinogenic and mutagenic action is evidence-based (WHO, 2000). Environmental contamination by PAHs, especially PAHs of high molecular mass, primarily refers to contamination of the atmosphere. PAHs with a smaller number of rings exist in the atmosphere in gaseous form, while those having a larger number of rings are adsorbed by air particles (Lee et al., 1981). Water and soil contamination by PAHs is regarded as secondary contamination since the air-borne PAHs are bound to suspended particles, and are deposited onto soil and water (Tritscher, 2004).

Table 6 and Fig. 8 show the PAHs status in soil in particular trial years. In the first trial year, the highest PAHs content was recorded in treatment II, and in treatments VII and VIII where crude oil was applied. Although the PAHs sum was almost equal in treatments II, VII and VIII, differences in particular PAHs are discernible between the treatments. In treatment II, due to its chemical composition and origin, benzo(b)fluorantene absolutely prevails in the PAHs sum, while all the 8 studied PAHs are equally represented in treatments VII and VIII where crude oil was applied (Table 6). This is attributed to the high volatility of PAHs (Samsøe-Petersen et al., 2002; Serrano et al., 2006). In treatments VII and VIII, where crude oil was applied, PAHs were not yet degraded by physical, chemical or biological processes. Owing to its total higher levels in soil compared to other PAHs, benzo(b)fluorantene remains longer in soil. At the time of repeated sampling in 2005, there were almost no PAHs in the soil; increased benzo(b)fluorantene was found only in treatment II (Table 6) compared to all the other treatments. At the last sampling in 2007, PAH presence in soil was insignificant because they were partially degraded by physical processes (Fismes et al., 2002 and

Samsøe-Petersen et al., 2002), by microorganisms (Samanta et al., 2002), or some other chemical or biological process (Riser-Roberts, 1998).

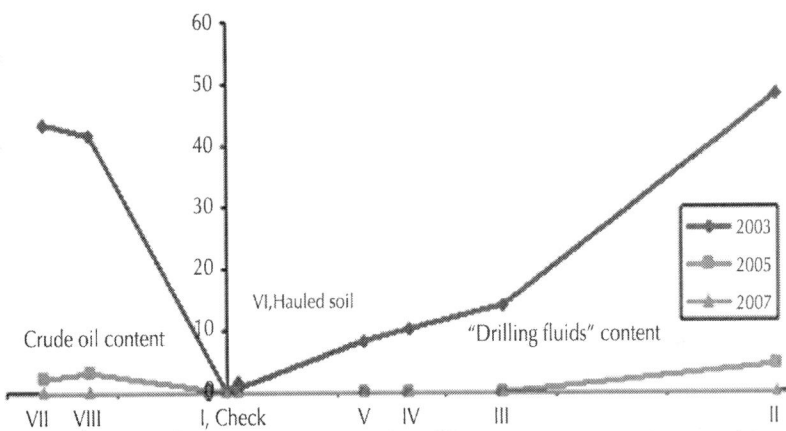

**Figure 8:** Effects of drilling fluids and crude oil on total Polycyclic aromatic hydrocarbons (PAHs) in soil, mg/kg.

## Heavy Metals

Of the numerous impacts on soil quality, a significant part is contributed by HM (Colins et al., 2006). This concept implies the presence of HM in soil in amounts causing a visible or measurable disturbance of some of the soil functions (Abrahams, 2002 and McLaren, 2003).

As soil is a very complex system encompassing numerous processes, some of which are irreversible and plants are subject to diverse biotic and abiotic influences, it is not always simple to assess soil and plant interactions and synergistic action, due to the occurrence of changes that would allow classifying the soil as contaminated. For these reasons, soil contamination is significantly different from (for example) air or water contamination. It is likewise very difficult to speak about threshold values of soil contamination by HM. Some values are applicable to heavy clay soils, whilst other values will hold for lighter sandy soils; some values will apply to soils used in agriculture, while other values will hold for soils located in various industrial or urban zones (Abrahams, 2002 and Samsøe-Petersen et al., 2002).

Table 7 show the status of soil contamination by certain HM (Cd, Hg, Pb, Mo, As, Ni, Co, Cu, Cr, Zn, Ba and V). The lowest HM content was in all cases determined in the control treatment. In all other treatments, changes in HM levels were conditioned by the degree of soil contamination. It can be seen from Table 7that cadmium, lead, molybdenum, nickel, cobalt and chromium contents did not differ significantly between the trial years. Mercury, arsenic, copper, zinc, barium and vanadium contents differed statistically between the trial years according to the degree of soil contamination. The highest mercury, arsenic, copper, zinc, barium and vanadium contents were determined in treatment II. This is associated with the chemical composition of drilling fluids applied in the trial (Table 2). As increased levels of mercury, arsenic, zinc and barium were found in these fluids, raised levels of the listed metals could be expected in treatments involving application of a certain percentage of drilling fluids. It is important to note that crude oil contains neither zinc nor barium; their increased presence in drilling fluids is attributed to the use of zinc carbonate and barite (barium sulphate) in oil drilling (Carls et al., 1995 and Scholten et al., 2000). Barium is insoluble, inert and non-toxic (Monaghan et al., 1980), so it is not considered a serious problem in soil. However, mercury, arsenic and zinc require quite a different approach. Higher levels of these metals in soil call for recultivation through electroremediation (Riser-Roberts, 1998), phytoremediation (Hazel, 2005) or solidification (Asia et al., 2007). Approximately equal copper and zinc contents were found in the treatment with drilling fluids, as well as in the treatment in which the contaminated soil was replaced by soil hauled to the site. This indicates the necessity to assess the quality of the soil to be transported to a pipeline breakage site.

**Table 7:** Total heavy metal contents in soil, mg/kg

| | Treatment year | I | II | III | IV | V | VI | VII | VIII |
|---|---|---|---|---|---|---|---|---|---|
| Cadmium | 2003 | 0.33 | 0.23 | 0.24 | 0.29 | 021 | 0.32 | 021 | 0.37 |
| | 2005 | 0.32 | 0.46 | 043 | 0.32 | 0.33 | 0.43 | 0.27 | 0.35 |
| | 2007 | 0.37 | 0.38 | 0.38 | 0.36 | 0.36 | 0.32 | 0.35 | 0.33 |
| Mercury | 2003 | 0.02 | 0.18** | 0.03* | 0.08** | 0.04** | 0.06** | 0.02 | 0.09** |
| | 2005 | 0.02 | 0.26** | 0.15** | 0.08** | 0.04** | 0.05** | 0.04** | 0.03* |
| | 2007 | 0.01 | 0.16** | 0.09** | 0.07** | 0.05** | 0.07** | 0.03* | 0.04** |
| Lead | 2003 | 13 | 18 | 14 | 19 | 15 | 14 | 14 | 12 |
| | 2005 | 13 | 18 | 18 | 15 | 16 | 15 | 20 | 15 |
| | 2007 | 15 | 26 | 21 | 20 | 20 | 31 | 17 | 19 |
| Molybdenum | 2003 | 0.5 | 0.4 | 0.4 | 0.5 | 0.7 | 0.6 | 0.4 | 0.4 |
| | 2005 | 0.6 | 0.4 | 0.6 | 0.6 | 0.4 | 0.8 | 0.4 | 0.6 |
| | 2007 | 0.7 | 1.0 | 1.1 | 1.1 | 1.0 | 1.1 | 1.0 | 1.0 |
| | 2003 | 7 | 18** | 14** | 16** | 9* | 9* | 9* | 10** |
| | 2005 | 6 | 32** | 16** | 19** | 11** | 11** | 10** | 8* |
| | 2007 | 7 | 33** | 15** | 16** | 9* | 31** | 7 | 8* |
| Nickel | 2003 | 17 | 25 | 23 | 28 | 20 | 20 | 20 | 19 |
| | 2005 | 17 | 15 | 18 | 15 | 15 | 15 | 17 | 14 |
| | 2007 | 12 | 14 | 14 | 13 | 13 | 14 | 12 | 12 |
| Cobalt | 2003 | 7 | 9 | 10 | 12 | 10 | 9 | 8 | 6 |
| | 2005 | 7 | 7 | 6 | 9 | 9 | 8 | 9 | 9 |
| | 2007 | 12 | 11 | 12 | 14 | 11 | 11 | 12 | 13 |
| Copper | 2003 | 15 | 18** | 16 | 15 | 12** | 18** | 15 | 14 |
| | 2005 | 14 | 20** | 18** | 14 | 12** | 19** | 14 | 10** |
| | 2007 | 14 | 22** | 18** | 17? | 15 | 21** | 14 | 14 |
| Chromium | 2003 | 40 | 30 | 25 | 55 | 25 | 50 | 22 | 22 |
| | 2005 | 40 | 45 | 26 | 27 | 24 | 30 | 25 | 16 |
| | 2007 | 30 | 45 | 38 | 36 | 33 | 44 | 31 | 31 |
| Zinc | 2003 | 70 | 78** | 65 | 69 | 70 | 67* | 55** | 63** |
| | 2005 | 70 | 81** | 61 | 65* | 70 | 71 | 50** | 45** |
| | 2007 | 51 | 92** | 74** | 70** | 62* | 91** | 54* | 58** |

| Barium | 2003 | 79 | 1872** | 1356** | 1423** | 709** | 195** | 419** | 97** |
|---|---|---|---|---|---|---|---|---|---|
| | 2005 | 75 | 1936** | 1228** | 1451** | 634** | 202** | 412** | 101** |
| | 2007 | 80 | 2000** | 1500** | 1500** | 800** | 500** | 400** | 101** |
| Vanadium | 2003 | 25 | 32** | 32** | 40** | 25 | 22** | 29** | 19** |
| | 2005 | 25 | 21** | 22** | 22** | 25 | 21** | 29** | 24 |
| | 2007 | 30 | 25** | 28** | 24** | 28* | 24** | 32* | 30 |

*,**Significant at the 0.05 and 0.01 levels of probability respectively.

## Changes in Plant Density and Yield of Crops Grown

In the first trial year, as early as in the germination/emergence stage, differences were observed in the emergence of winter wheat (cultivar: Zlatni dukat) depending on the different TPH levels in the soil. Winter wheat emergence was inversely proportional to TPH levels in the soil. In the treatments with higher soil contamination (treatments II; IV; VII and VIII), crop emergence was much poorer compared to treatments III and V with lower TPH content. Higher contamination caused formation of a thin film around the seed germ and thereby prevented the inflow of oxygen, which caused embryo death. Another reason for poorer emergence is the fact that TPH contaminated soil becomes more compact and less moist, and has a higher content of toxic substances. Ferrell et al. (1984), Issoufi et al. (2006), Shahriari et al. (2007), Adedokun and Ataga (2007) also report that oil pollution inhibits seed germination and plant growth. If the number of plants determined in the control treatment (100%) is taken as optimal, it can be seen that plant density at the wheat harvest in treatment II was 74% (that) of the control, in treatment III — 98%, in treatment IV — 60%, in treatment V — 88%, while the plant density in treatment VI was only 50% (that) of the control (Table 8). Plant density in treatments VII and VIII was 57% and 37% (that) of the control, respectively. This indicates that crude oil had a much stronger effect on the achieved plant density than the drilling fluids. Plant density achieved in the control treatment was statistically higher compared to other treatments, whereas no statistically significant differences were determined between other treatments (Table 8).

**Table 8:** Yield, plant density and some components of winter wheat

| Treatment/year/components | I | II | III | IV | V | VI | VII | VIII |
|---|---|---|---|---|---|---|---|---|
| 2003/04 | | | | | | | | |
| Yield, g pot$^{-1}$ | 44.5 | 22.3** | 31.7** | 26.9** | 28.0** | 26.9** | 22.4** | 23.6** |
| Total plants | 106 | 79 | 104 | 63** | 93 | 53** | 60** | 39** |
| Plants with ears | 71 | 63 | 74 | 47** | 60 | 43** | 48** | 37** |
| Plants without ears | 35 | 16** | 30 | 16** | 33 | 10** | 12** | 2** |
| 2005/06 | | | | | | | | |
| Yield, g pot$^{-1}$ | 46.6 | 42.6 | 49.0 | 45.6 | 41.6 | 48.1 | 36.4* | 37.8* |
| Total plants | 129 | 86* | 114 | 123 | 113 | 112 | 115 | 115 |
| Plants with ears | 108 | 81* | 96 | 103 | 95 | 94 | 96 | 97 |
| Plants without ears | 21 | 5** | 18** | 20* | 18** | 18** | 19** | 18 |

*,**Significant at the 0.05 and 0.01 levels of probability respectively.

According to the crop sequence, Winter barley, cv. Rex was sown into pots in the second trial year. Even at the initial emergence stages, differences from the first trial year (when wheat was sown) were observable. The number of plants and the plant density achieved did not reflect the differences that were recorded during the first year. No statistically significant differences between trial treatments were determined with respect to the number of plants. The achieved yield, however, shows statistical differences, {but not as marked as in the first year} (Table 9). The highest difference was found between treatments I and II, and treatment VIII, while other treatments had more or less uniform yields.

**Table 9:** Yield, plant density and some components of winter barley

| Treatment/year/components | I | II | III | IV | V | VI | VII | VIII |
|---|---|---|---|---|---|---|---|---|
| 2004/05 | | | | | | | | |

| Treatment/year/components | I | II | III | IV | V | VI | VII | VIII |
|---|---|---|---|---|---|---|---|---|
| 2005 | | | | | | | | |
| Yield, g pot$^{-1}$ | 32.7 | 18.0** | 19.2** | 17.9** | 20.3* | 4.9** | 23.2* | 22.4* |
| Yield, g pot$^{-1}$ | 61.3 | 63.9 | 56.2* | 54.2* | 57.1 | 57.3 | 54.9 | 49.6** |
| Total plants | 119 | 128 | 118 | 106 | 109 | 122 | 117 | 131 |
| Plants with ears | 114 | 128* | 116 | 106 | 108 | 119 | 111 | 100* |
| Plants without ears | 4.5 | 0** | 2** | 0** | 1** | 3** | 6** | 31** |
| 2006/07 | | | | | | | | |
| Yield, g pot$^{-1}$ | 63.8 | 61.8 | 56.2* | 54.2* | 57.1* | 57.4 | 55.0* | 54.7* |
| Total plants | 129 | 128 | 118 | 110 | 112 | 122 | 117 | 116 |
| Plants with ears | 124 | 124 | 116 | 106 | 109* | 119 | 111 | 109* |
| Plants without ears | 5 | 4** | 2** | 3** | 3** | 3** | 6** | 6** |

*,**Significant at the 0.05 and 0.01 levels of probability respectively.

After the barley was harvested, soybean, cv. Sabina (00-000 maturity group) was sown into pots. The achieved plant density and yield manifested very interesting changes compared to the preceding year when barley was grown. As usual, uniform seed rate was applied to all pots. However, essential differences in plant density and yield were already observed at the emergence stage, as well as in the growing period. The largest number of plants was recorded in the control treatment over the entire growing period, while the number of plants differed statistically in other treatments (Table 10). By far the lowest plant density was achieved in treatment VI, while that of other treatments was significantly lower than the control. The highest yield was obtained in the control treatment, while significantly lower yields were recorded in other treatments. Although three years had passed since the beginning of the trial, it was evident that TPH contamination exerted a significant effect on the number of emerged plants, and thereby also on the achieved soybean density.

**Table 10:** Yield, plant density and some components of Soybean

| Total plants | 14 | 8** | 6** | 5** | 9** | 2** | 8** | 8** |
| Total pods | 104 | 54** | 51** | 44** | 73 | 13** | 76 | 72 |
| 2006 | | | | | | | | |
| Yield, g pot$^{-1}$ | 14.6 | 11.1 | 14.3 | 13.4 | 11.6 | 15.9 | 16.7 | 16.3 |
| Total plants | 8 | 8 | 9 | 8 | 8 | 8 | 8 | 9 |
| Total pods | 37 | 35 | 39 | 34 | 40 | 46** | 47** | 53** |

*,**Significant at the 0.05 and 0.01 levels of probability respectively.

Winter wheat (the same cultivar as in the first year) was sown again in the autumn of 2005. Early in the initial emergence stages, changes in plant density were observed in comparison with the first trial year. Like in the first year, about 200 germinated seeds were sown per pot. Plant density in treatment II was 70% and in treatment IV 95% that of the control treatment. Plant density of other treatments was about 90% (that) of the control.

Soybean, this time cultivar Dora (0-00 maturity group), was sown again in July 2006. Compared to 2005, when cultivar Sabina was sown in pots, plant density was very different in 2006. This calls for an explanation of the difference in the number of plants that emerged in the control treatment in the two years when soybean was grown in pots (Table 10). Sabina is a soybean cultivar that is sown at a higher density (1,400,000 plants per hectare), whereas the sowing density of cultivar Dora is 650,000 plants per hectare. This accounts for the different number of plants that emerged in particular treatments. No statistically significant differences in the number of emerged plants were recorded in 2006, while the total number of pods in treatments VI, VII and VIII was statistically higher compared to the control. No statistically significant differences in yield were found between the studied treatments.

Upon soybean harvest in the autumn of 2006, winter barley was sown into pots, the same cultivar as in 2004. Data given in Table 9 show an approximately equal plant density at earing and at harvest in all treatments. There was no statistically significant difference between the control and treatments II and VI either. Significant differences, however, were recorded relative to the other treatments. Similar conclusions regarding plant density and yields achieved were reached by Shahriari et al. (2007), Akaninwor et al. (2007). They determined differences in yield and plant density between stubble crops and soybean as a spring row crop. Chaineau et al. (2003) report that the resistance of seeds to oil contamination followed the decreasing order: wheat > barley > maize > pea > lettuce.

The presented results of the four-year investigations and growing six crops indicate that the application of drilling fluids and crude oil leads to changes both in soil and in crops grown. These changes were most evident in the first trial year, with a marked decline in some parameters in the subsequent years. Upon application of the studied materials, soil pH, OM and HM contents remained constant throughout the trial period, whereas significant changes occurred in the case of TPH, MO and PAHs after the first trial year. As drilling fluids are rich in calcium and carbon, (only if contaminants (HM) were removed from drilling fluids) these materials could be used for liming acid soils as soil improvers only if contaminants (HM) were removed from them (drilling fluids) (Miller and Pesaran, 1980).

## CONCLUSIONS

(I would write this in present SK) drilling fluids have (had) a stronger influence on the studied soil chemical properties while crude oil affected plant density and crop yield more strongly. The level of soil contamination by TPH and PAHs in the first trial year had a crucial role for the achieved plant density and yield. The raised content of TPH and PAHs in soil has the strongest influence in the emergence stage. TPH and PAHs in soil forms a thin film around the seed germ and thereby hampers oxygen inflow, leading to embryo death and/or slower emergence of the plant.

TPH levels below 5 g/kg soil or 5 mg/kg PAHs in soil had no significant effect on the plant density of crops grown. Consequently, the value of 5 g/kg TPH and 5 mg/kg PAHs in soil could be recommended as a warning or emergency value in remediation of hydrocarbons-contaminated soil. The presented results show that the largest part of TPH and PAHs were lost through bioremediation (bioaugmentation and/or biostimulation) processes and soil mixing during the first and second trial years. On the other hand, the levels of heavy metals did not change much over the four years trial. This indicates that the problem of soil contamination by TPH, and partially the PAHs problem, may be solved through soil bioremediation or aeration by tilling practices (ploughing and harrowing). Increased content of heavy metals in soil requires quite a different approach to soil cleansing by electroremediation, phytoremediation or solidification.

The extent of environmental consequences and changes in the soil chemical complex, plant density and yield of crops grown on TPH and PAHs contaminated soil depend primarily on the type and quantity of crude oil or drilling fluids unintentionally introduced into the environment, soil type, crops grown, the season, climate, and various other influences.

# ACKNOWLEDGEMENTS

This paper presents results of research programs supported by the Ministry of Science, Education and Sports of Republic Croatia.

# REFERENCES

1. Abrahams, P.W., 2002. Soils: their implications to human health. The Science of The Total Environment 291 (1–3), 1–32.
2. Adam, G., Duncan, H., 2002. Influence of diesel fuel on seed germination. Environmental Pollution 120 (2), 363–370.
3. Adedokun, O.M., Ataga, A.E., 2007. Effects of amendments and bioaugumentation of soil polluted with crude oil, automotive gasoline oil, and spent engine oil on the growth of cowpea (Vigna ungiculata L. Walp). Scientific Research and Essay 2 (5), 147–149.

4. Agbogidi, O.M., Eruotor, P.G., Akparobi, S.O., Nnaji, G.U., 2007. Evaluation of crude oil contaminated soil on the mineral nutrient elements of maize (Zea Mays L.). Journal of Agronomy 6 (1), 188–193.
5. Akaninwor, J.O., Ayeleso, A.O., Monago, C.C., 2007. Effect of different concentrations of crude oil (Bonny light) on major food reserves in guinea corn during germination and growth. Scientific Research and Essay 2 (4), 127–131.
6. Asia, I.O., Jegede, S.I., Jegede, D.A., Ize-Iyamu, O.K., Akpasubi, E.B., 2007. The effects of petroleum exploration and production operations on the heavy metals contents of soil and groundwater in the Niger Delta. Journal of Physical Science 2 (10), 271–275.
7. Baker, J.M., 1970. The effects of oils on plants. Environmental Pollution 1 (1), 27–44.
8. Bauder, T.A., Barbarick, K.A., Ippolito, J.A., Shanahan, J.F., Ayers, P.D., 2005. Soil properties affecting wheat yields following drilling-fluid application. Journal of Environmental Quality 34, 1687–1696.
9. Callaham, M.A., Stewart, A.J., Alarcon, C., McMillen, S.J., 2002. Effects of earthworm (Eisenia fetida) and wheat (Triticum aestivum) straw additions on selected properties of petroleum contaminated soils. Environmental Toxicology and Chemistry 21/8, 1658–1663.
10. Carls, E.G., Dennis, B.F., Chaffey, S.A., 1995. Soil contamination by oil and gas drilling and production operations in Padre Island National Seashore, Texas, USA. Journal of Environmental Managment 45, 273–286.
11. Carr, R.H., 1919. Vegetative growth in soils containing crude petroleum. Soil Science 8, 67–68.
12. Chaineau, C.H., Yepremian, C., Vidalie, J.F., Ducreux, J., Ballerini, D., 2003. Bioremediation of a crude oil-polluted soil. Biodegradation, Leaching and Toxicity Assesments 144, 419–440.

13. Colins, C., Fryer, M., Grosso, A., 2006. Plant uptake of non-ionic organic chemicals. Environmental Science and Technology 40, 45–52.
14. De Jong, E., 1980. The effect of a crude oil spill on cereals. Environmental Pollution 22, 187–196.
15. Dorn, P.B., Salanitro, J.P., 2000. Temporal ecological assesment of oil contaminated soils before and after bioremediation. Chemosphere 40, 419–426.
16. Ekundayo, E.O., Emede, T.O., Osayande, D.I., 2001. Effects of crude oil spillage on growth and yield of maize (Zea mays L.) in soils of midwestern Nigeria. Plant Foods for Human Nutrition 56 (4), 313–324.
17. Ferrell, R.E., Seneca, E.D., Linthurst, R.A., 1984. The effects of crude oil on the growth of Spartina alterniflora Loisel and Spartina cynosuroides (L.) Roth. Journal of Experimental Marine Biology and Ecology 83 (1), 27–39.
18. Fismes, J., Perrin-Ganier, C., Empereur-Bissonet, P., Morel, J.L., 2002. Soil-to-root transfer and translocation of polycyclic aromatic hydrocarbons by vegetables grown on industrial contaminated soils. Journal of Environmental Quality 31, 1649–1656.
19. Hazel, W., 2005. Suck it up. Phytoremediation. Available at http://ourgardengang.tripod.com.
20. Healy, M., Wise, D.L., Moo-Young, M., 2001. Environmental Monitoring and Biodiagnostics of Hazardous Contaminants. Kluwer Academic Publishers, Boston-London.
21. Issoufi, I., Rhykerd, R.L., Smiciklas, K.D., 2006. Seedling growth of agronomic crops in crude oil contaminated soil. Journal of Agronomy and Crop Science 192, 310–317.
22. Jobson, A., McLaughlin, M., Cook, F.D., Wstlake, D.W.S., 1974. Effects of amendments on the microbial utilisation of oil applied to soil. Journal of Applied Microbiology 27, 166–171.
23. Kelsey, J.W., Alexander, M., 1997. Declining bioavalibility and inappropriate estimation of risk of persistent compounds.

Environmental Toxicology and Chemistry 16, 582–585.
24. Kirkpatrick, W.D., White Jr., P.M., Wolf, D.C., Thoma, G.J., Reynolds, C.M., 2006. Selecting plants and nitrogen rates to vegetate crude-oil-contaminated soil. International Journal of Phytoremediation 8 (4), 285–297.
25. Kisic, I., Basic, F., Mesic, M., Veronek, B., Vadjic, Z., Mesic, S., 2005. Changes in soil and crop yield caused by oil incidents. Cereal Research Communications 33 (1), 243–246.
26. Kyung-Hwa, B., Hee-Sik, K., Hee-Mock, O., Byung-Dae, Y., Jaisoo, K., In-Sook, L., 2004. Effect of crude oil, oil components, and bioremediation on plant growth. Journal of Environmental Science and Health A39 (9), 2465–2472.
27. Lee, M.L., Novotny, M.V., Bartle, K.D., 1981. Analytical chemistry of Polycyclic Aromatic Compounds. Academic Press, New York.
28. Mackin, J.G., 1950. Report on a study of the effects of application of crude petroleum on saltgrass (Distichlis spicata L.) Green. Texas A&M Research Foundation.
29. McLaren, R.G., 2003. Micronutrients and toxic elements. In: Benbi, D.K., Nieder, R. (Eds.), Handbook of processes and modeling in the Soil-Plant System. The Haworth Press, New York, pp. 589–618.
30. Mendelssohn, I.A., Hester, M.W., Sasser, C., Fishel, M., 1990. The effect of a Lousiana crude oil discharge from a pipeline break on the vegetation of a Southeast Lousiana brackish marsh. Oil and chemical pollution 7 (1), 1–15.
31. Miller, R.W., Pesaran, P., 1980. Effects of drilling fluids on soil and plants. II Complete drilling fluid mixtures. Journal of Environmental Quality 9, 552–556.
32. Monaghan, P.H., McAuliffe, C.D., Weis, 1980. Environmental aspects of drilling fluids and cuttings from oil and gas operation in offshore and coastal waters. Marine Environmental Pollution 1: Hydrocarbons. Elsevier Scientific.

33. Murphy, J.F., Riley, J.P., 1929. Some effects on crude petroleum on nitrate production, seed germination and growth. Soil Science 24, 117–120.
34. Ogboghodo, I.A., Iruaga, E.K., Osemwota, I.O., Chokor, J.U., 2004a. An assessment of the effects of crude oil pollution on soil properties, germination and growth of maize (Zea Mays) using two crude types — Forcados Light and Escravos Light. Environmental Monitoring and Assessment 96, 143–152.
35. Ogboghodo, I.A., Erebor, E.B., Osemwota, I.O., Isitekhale, H.H., 2004b. The effects of application of poultry manure to crude oil polluted soils on maize growth and soil properties. Environmental Monitoring and Assessement 96 (1–3), 153–161.
36. Okolo, J.C., Amadi, E.N., Odu, C.T.I., 2005. Effects of soil treatments containing poultry manure on crude oil degradation in a sandy loam soil. Applied Ecology and Environmental Research 3 (1), 47–53.
37. Pezeshki, S.R., Hester, M.W., Lin, Q., Nyman, J.A., 2000. The effects of oil spill and cleanup on dominant US Gulf coast marsh macrophytes: a review. Environmental Pollution 108 (2), 129–139.
38. Rhykerd, R.L., Crews, B., McInnes, K.J., Weaver, R.W., 1999. Impact of bulking agents, forced aeration, and tillage on remediation of oil-contaminated soil. Bioresource Technology 67 (3), 279–285.
39. Riser-Roberts, E., 1998. Biodegradation/mineralization/biotransformation/bioaccumulation of petroleum. Remediation of petroleum contaminated soils — Biological, Physical and Chemical Processes. Lewis Publishers, Florida, pp. 115–192.
40. Samanta, S.K., Singh, O.V., Jain, R.K., 2002. Polycyclic aromatic hydrocarbons: environmental pollution and bioremediation. Trends in Biotechnology 20 (6), 243–248.
41. Samsøe-Petersen, L., Larsen, H.E., Larsen, B.P., Brun, P., 2002. Uptake of trace elements and PAHs by fruit and

vegetables from contaminated soils. Environmental Science and Technology 36, 3057–3063.
42. Sarkar, D., Ferguson, M., Datta, R., Birnbaum, S., 2005. Bioremediation of petroleum hydrocarbons in contaminated soils: comparison of biosolids addition, carbon supplementation, and monitored natural attenuation. Environmental Pollution 136 (1), 187–195.
43. Scholten, M.C.Th., Karman, C.C., Huwer, S., 2000. Exotoxicological risk assesment related to chemicals and pollutants in off-shore oil production. Toxicology Letters 112–113, 283–288.
44. Serrano, A., Gallego, M., González, J.L., 2006. Assessment of natural attenuation of volatile aromatic hydrocarbons in agricultural soil contaminated with diesel fuel. Environmental Pollution 144 (1), 203–209.
45. Shahriari, M.H., Savaghebi-Firoozabadi, G., Azizi, M., Kalantari, F., Minai-Tehrani, D., 2007. Study of growth and germination of Medicago Sativa (Alfalfa) in light crude oil-contaminated soil. Research Journal of Agriculture and Biological Sciences 3 (1), 46–51.
46. Tritscher, A.M., 2004. Human health risk assessment of processing-related compounds in food. Toxicology Letters 149, 177–186.
47. WHO, 2000. Polycyclic Arpomatic Hydrocarbons (PAHs). Air Quality Guidelines. WHO, Copenhagen, Denmark.
48. Yong, R., Mohamed, A.M.O., Warkentin, 1992. Principles of Contaminant Transport in Soils. Elsevier Science Publishers, Amsterdam, p. 328.

# Chapter 9

# Influence of Viscosity Modifier Nature and Concentration on the Viscous Flow Behaviour of Oil-Based Drilling Fluids at High Pressure

Ivica Kisic[a], Sanja Mesic[b], Ferdo Basic[a], Vladislav Brkic[b], Milan Mesic[a], Goran Durn[c], Zeljka Zgorelec[a], Lidija Bertovic[b]

Departamento de Ingeniería Química, Centro de Investigación de Tecnología de Productos y Procesos Químicos (Pro$^2$TecS), Universidad de Huelva, Facultad de Ciencias Experimentales, Campus del Carmen, 21071 Huelva, Spain

## ABSTRACT

This work deals with the effect of viscosity modifier nature and concentration on the rheological properties of model oil-based drilling fluids (OBM) submitted to high pressure. The oil-based fluids were formulated by dispersing, with a high shear mixer, two selected organobentonites in a mineral oil, at room temperature. The viscous flow behaviour of the corresponding dispersions was characterised as a function of pressure, organoclay nature and organoclay concentration, using a controlled-stress rheometer equipped with both pressure cell and coaxial cylinder geometries. A factorial Sisko–Barus model, which takes into account both shear and pressure effects in the same equation, fitted the experimental pressure–viscosity data fairly well.

The influence of disperse phase concentration on the shear-thinning characteristics of these organoclay dispersions is related to the development of different microstructures, which depend on organoclay nature. In this sense, the resulting microstructure has been attributed to the cohesion energy between microgels domains. From the experimental results obtained, it can be concluded that the viscous flow behaviour of the OBM investigated is strongly affected by organoclay nature and concentration. The pressure–viscosity behaviour of these dispersions is mainly influenced by the piezoviscous properties of the oil and the properties of the continuous phase. The Sisko–Barus model proposed can be a useful tool, from an engineering point of view, for calculating pressure losses in the different sections of the bore, as well as being of significant help to solve other additional problems, such as hole cleaning, induced fracturing, and hole erosion during the drilling operation.

## INTRODUCTION

Oil-based drilling fluids called oil based muds (OBM) are dispersions usually showing a complex rheology. Regarding the nature of their continuous phase, fluids used in drilling and completion wells can be classified into two main groups, water-based and oil-based. The main functions of these fluids are: i. To carry cuttings from the bottom of the hole, transport them up and remove rock bit at the surface. ii. To cool and clean the drill and the bit. iii. To maintain the stability of borehole. iv. To lubricate the gap between the drilling string and the wall of the hole. v. To prevent the inflow of fluids from surrounded rocks. vi. To form a thin and low-permeable filter cake. vii. To be non-damaging to the producing formation. viii. To be non-hazardous to the environment and personnel (Chilingarian and Vorabutr, 1983 and Menezes et al., 2010).

Most of global drilling operations use water-based muds (WBM), because of their lower environmental impact, whereas only 5–10% of the wells drilled use OBM (Caenn and Chillingar, 1996 and Meng et al., 2012). Nevertheless, OBM have interesting features to overcome certain undesirable characteristics of the water-based ones, such as better lubrication and higher boiling points (Khodja et al., 2010). Basically, OBM can be classified into three categories: (1) All-oil muds, consisting of a mixture of organoclay (OC) and synthetic or mineral oil, which are used for minimum pressure losses and low permeability reservoirs; (2) Oil muds, consisting of OC, emulsifiers, oil and water (2–10 m%), which are designed for well stabilisation at high temperature; (3) Invert oil muds, consisting of OC, emulsifiers, oil, additives and water (up to 40 m%), which are used for shale stability and improved penetration.

The relationship between flow behaviour and composition is an important issue to formulate suitable OBM. Regarding composition, clays are a key component for developing some specific properties of these dispersions, which will be submitted to the extreme pressure and temperature conditions of the wellbore. Concerning OBM, organophilic bentonites have been extensively used due to both good dispersing properties in the oil phase and filtration characteristics (Jordan et al., 1965). These OC result from the reaction of smectite-type and amine cationic groups, without addition of supplementary additives (Hauser, 1950 and Jordan, 1949).

Regarding flow properties, OBM are frequently submitted to extreme shear, temperature and pressure conditions in downhole operations. During the circulation of the drilling fluid around the wellbore, the shear rate may vary from zero to more than 1000 $s^{-1}$, whilst temperature can vary from values below 5 °C in water settings to above 200 °C at the bottom during the round trip. In addition, the pressure exerted by the mud column may be as much as 1400 bar at the deepest part (Darley and Gray, 1988). These severe conditions may change the bulk rheological behaviour of the dispersions because of pressure–temperature-dependent viscosity changes and particle–particle interactions modifications (Briscoe et al., 1994).

Several authors have examined the evolution of WBM viscosity with pressure and temperature, temperature being the most important factor (Santoyo et al., 2001 and Wang et al., 2010). Furthermore, other studies have been mainly focused on the use of new additives as rheology-modifiers to improve drilling operation. With this aim, recent studies have concentrated their efforts on how to overcome the hole instabilities, related to the aqueous media at extreme conditions, by using different additives, such as viscoelastic surfactants or synthetic polymers (England and Parris, 2010 and Wang et al., 2011).

For OBM, studies concerning the effect that pressure exerts on both their rheological behaviour and physical properties are very scarce. Combs and Whitmire (1968) studied the effect of temperature and pressure on the rheology of OBM formulated with OC, and found that the change in continuous phase viscosity was the main controlling factor. Politte (1985) concluded that the plastic viscosity could be normalised using the viscosity of the oil medium, whereas the yield stress is a weak function of pressure. Besides, Houwen and Geehan (1986) found a simple model to determine both yield-stress and high-shear-rate viscosity of invert muds as a function of pressure and temperature, using up to four parameters. In most cases, changes observed in physical properties and flow behaviour of OBM have been explained on the basis of the effect that both temperature and pressure exert on the viscosity of the continuous phase (Gandelman et al., 2007 and Herzhaft et al., 2001). Much less attention has been devoted to the effect that nature and concentration of viscosity modifiers exert on the rheological properties of oil dispersions submitted to high pressure (Ghalambor et al., 2008), probably due to the experimental constraints involving high pressure rheology measurements with fluids that exhibit non-Newtonian behaviour. Consequently, the overall objective of this work was to study the effect that viscosity modifier nature and concentration exert on the rheological properties of model OBM submitted to high pressure.

# EXPERIMENTAL

## Materials

Two commercially available OC, denoted as B34 and B128 and provided by Elementis (Belgium), were used in the present study. Their chemical formula and some physical characteristics are shown in Table 1.

**Table 1:** OC used in this study

| Commercial name | Clay mineral | Abbreviated notation for intercalated ions[a] | Chemical formula | $d_{001}$[b] (nm) |
|---|---|---|---|---|
| Bentone 34 | Bentonite | 2M2HT | HT—N⁺(CH₃)(CH₃)—HT | 2.767 |
| Bentone 128 | Bentonite | 2MBHT | CH₃—N⁺(CH₃)(HT)—CH₂—⌬ | 3.344 |

[a]The abbreviations of quaternary ammonium ions corresponds to: M: methyl, B: benzyl, HT: hydrogenated tallow.

[b]Basal spacing determined by X-ray diffraction (XRD).

A mineral based lubricating oil, SR-10 (916 kg/m$^3$ and 115 cSt, at 40 °C) supplied by Verkol (Spain), was used as base oil for the formulation of OBM.

## Samples Preparation

Organobentonite dispersions were prepared by mixing OC (at concentration of 1, 3 and 5 m%.) in SR-10 oil base, at room temperature, using a high mixer Ultraturrax (Ika, Germany), at a rotational speed of 9000 rpm for five minutes. Prior to high shear processing, the OC were wetted with the oil, at room temperature, in a low shear mixer using a conventional four blade impeller.

## X-Ray Diffraction (XRD)

XRD measurements were carried out on OC powders and their oily dispersions, at room temperature, using a Bruker S8 Advance (Germany) diffractometer equipped with a secondary monochromator, a Brentano Bragg geometry goniometer and a copper cathode as X-ray source. The samples were subjected to Cu Kα radiation with a wavelength of 0.15406 nm. The 2θ angles varied from 1.5° to 20°, single scanning step of 0.017°, and measurement time of 6 s per step. The high intensity peaks in XRD curves show the $d_{001}$-spacing, which have been included in the Bragg equation to determine interlayer distance of each organically modified bentonite.

## Optical Microscopy

Optical microscopy observations were carried out by using an Olympus BX52 (Japan) microscope, equipped with an Olympus C5050Z camera and an objective of 20× and 50×. An electric heating system LTS 350 (Linkam, UK) coupled with microscope stand was used to maintain the temperature constant. The OC dispersions were carefully poured into a sample holder and spread under the glass cover slip at room temperature. Before observations, all samples were heated up to 40 °C to compare both optical and rheological results.

## Viscous Flow Measurements

Viscous flow measurements were performed using a controlled-stress rheometer, MARS II from Thermo-Scientific (Germany). Rheological data were obtained using a coaxial cylinder geometry (41 mm inner diameter, 1 mm gap, 60 mm length) at atmospheric pressure, and a coaxial cylinder-pressure cell D400/200 at high pressure. The cell D400/200 is a pressure vessel of 39 mm of inner diameter. Inside the cell, an inner cylinder of 38 mm diameter and 80 mm length was put in contact with a sapphire surface at the bottom of the vessel by a steel needle. This inner cylinder was equipped, at the top, with a secondary magnetic cylinder (36 mm diameter, 8 mm length), magnetically coupled to a tool outside the cell, which was connected to the motor-transducer of the rheometer. The pressure cell was connected to a hydraulic pressurisation system through a needle control valve.

A pressure transducer GMH 3110 (Gresingeg Electronic, Germany), able to measure differential pressures ranging 0 to 400 bar (0.1 bar resolution), was used.

Both atmospheric and high pressure rheological measurements were performed at 40 ± 0.1 °C using a circulating silicone bath.

Steady-state flow curves were obtained without sample pre-shear. The measurements were carried out by applying an increasing shear rate ramp, in the range comprised between 0.01 and 1000 $s^{-1}$. Two replicates of each rheological test were performed on fresh samples. The experimental error in viscosity was always inferior to ± 5%. Due to the fact that the viscous flow measurements might be influenced by a sedimentation process during the experimental flow time (Tropea et al., 2007), sedimentation rates were tested, at the above-mentioned temperature, to ensure that the flow measurements were not affected by phase separation.

# RESULTS AND DISCUSSION

## Viscous Flow Behaviour of OBM: Effect of OC Nature and Concentration

Viscous flow curves for the base oil and the drilling oil dispersions formulated with B34 and B128 OC, respectively, are shown in Fig. 1 and Fig. 2. As can be observed, OC dispersions (1–5 m %) display a shear-thinning behaviour with a tendency to reach a high-shear-rate-limiting viscosity. This shear-thinning behaviour is more apparent as OC concentration increases.

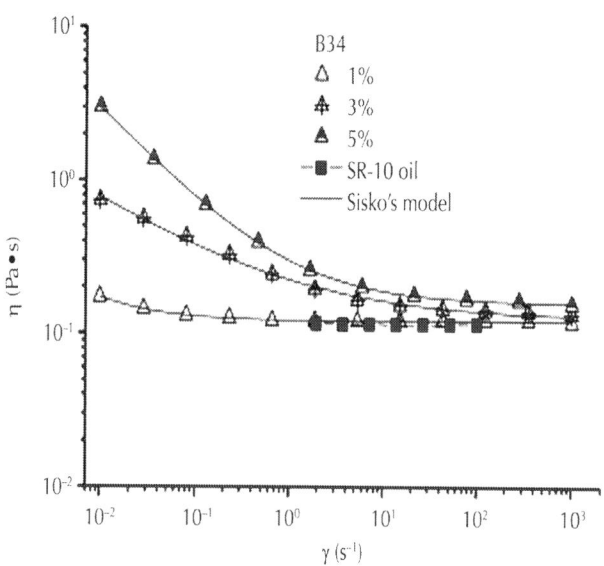

**Figure 1:** Viscous flow curves of OBM as a function of B34 OC concentration, at atmospheric pressure and 40 °C.

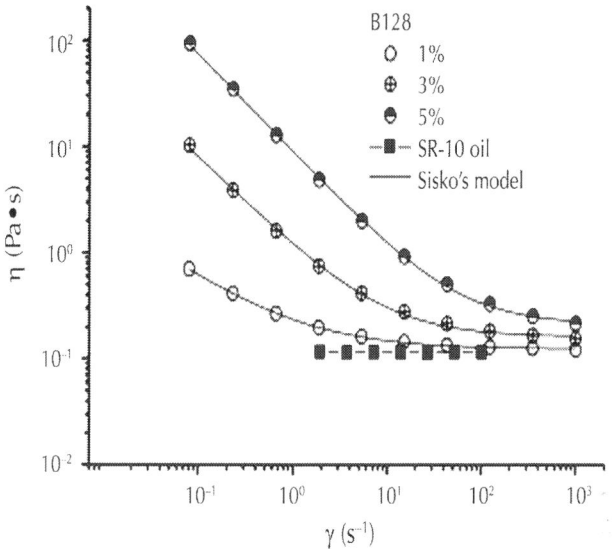

**Figure 2:** Viscous flow curves of OBM as a function of B128 OC concentration, at atmospheric pressure and 40 °C.

As can be seen in Fig. 1 and Fig. 2, Sisko's model (Turian et al., 1998, Turian et al., 2002 and Weir and Bailey, 1996) fits the viscous flow behaviour of these OC (1–5%) dispersions fairly well:

$$\eta = \eta_{\infty 0} + k_0 \dot{\gamma}^{n_0 - 1} \tag{1}$$

Where $\eta$ is the apparent viscosity, $\eta_{\infty 0}$ is the high-shear-rate-limiting viscosity, $k_o$ the consistency index, and $n_o$ the flow index. Sisko's parameter values are shown in Table 2. The values of the average absolute relative deviation (%AARD), which is defined as:

$$\%AARD = \frac{100}{n} \sum_{i=1}^{n} \left| \frac{\eta_{i,\exp} - \eta_{i,cal}}{\eta_{i,cal}} \right| \tag{2}$$

are also shown in Table 2.

**Table 2:** Some rheological parameters of the drilling fluids studied, as a function of OC concentration and pressure

| % wt. | $k = k_o + k_1 \Delta P$ | | $n = n_o + n_1 \Delta P$ | | $\eta_\infty$ (Pa s) | $\beta$ (bar$^{-1}$) | AARD (%) |
|---|---|---|---|---|---|---|---|
| | $k_o$ (Pa s$^n$) | $k_1$ (Pa s$^n$ bar$^{-1}$) | $n_o$ | $n_1$ (bar$^{-1}$) | | | |
| **B128** | | | | | | | |
| 1 | 0.17 | −6.9E−5 | 0.65 | 1.2E−5 | 0.100 | 0.0027 | 10.1 |
| 3 | 1.27 | −6.8E−4 | 0.16 | 3.4E−5 | 0.153 | 0.0027 | 5.11 |
| 5 | 9.55 | −1.1E−2 | 0.09 | 5.6E−5 | 0.195 | 0.0028 | 3.03 |
| **B34** | | | | | | | |
| 1 | 0.0081 | 1.0E−4 | 0.65 | −9.5E−4 | 0.118 | 0.0027 | 1.77 |
| 3 | 0.097 | 6.3E−4 | 0.55 | −9.4E−4 | 0.129 | 0.0027 | 11.6 |
| 5 | 0.14 | 4.7E−4 | 0.36 | −3.9E−4 | 0.163 | 0.0027 | 6.01 |
| **Oil** | | | | | | | |
| | | | | | 0.114 | 0.0027 | |

The shear-thinning behaviour, observed for these dispersions, can be explained as a particular characteristic of elastic soft solids (King et al., 2007), especially in the case of concentrated dispersions. Thus, the rheological behaviour of these materials has been related to both dispersion state and volume fraction of the disperse phase, which could be modified by physical and/or chemical interactions between the OC molecules and oil medium. In this sense, Moraru (2001) investigated the influence that the type of OC exerts on the shear-thinning characteristics of non-aqueous media. Primarily, the rheological behaviour of these non-polar colloidal systems was affected by its morphology, related to different aggregation states of the clay from nano to macroscale. Taking into account that organobentonites are clays partially covered by alkylammonium molecules adsorbed at their surface, the structure and, consequently, the flow behaviour of these dispersions may be related to the interactions developed between the organophilic ions and the solvent, the organic chain density between platelets and the chemical nature of the medium (Le Pluart et al., 2004). These interactions, which normally increase with clay concentration, lead to an increase in viscosity, as has been pointed out elsewhere (Das Kanungo and McAtee, 1986).

In this sense, a remarkable increase in viscosity with clay concentration is shown in Fig. 1 and Fig. 2, for both OC, more important for B128 dispersions, indicating that an increase in shear rate could yield a progressive alignment of the platelets, leading to a pronounced shear-thinning behaviour (Massinga et al., 2010).

The different viscosity values observed in the low shear rate region for dispersions of similar concentration (higher values for B128 dispersions) suggest that the microstructure developed in both OC should be completely different. Previous results pointed out that the decrease in the flow index could be related to a higher degree of intercalation in clay platelets (Wagener and Reisinger, 2003). According to this, the flow behaviour observed for B128 dispersions, which present lower flow indexes (0.09–0.64) than B34 dispersions, seems to indicate that oil molecules easily penetrate into the interlayer space of the B128 OC, yielding a stronger structural network for this OC (Hato et al., 2011 and Zhang et al., 2003). In addition, it can be deduced from Fig. 1 and Fig. 2 that, for B34 dispersions, the high-shear-rate-limiting viscosity is achieved at lower shear rate, indicating the development of a weaker structure.

As has been previously reported, the rheology of dispersion is a complex function of several factors, such as disperse phase volume fraction, particle shape and particle interactions (Mueller et al., 2010). At high shear rates, the viscosity values of the dispersions studied in this research mainly depend on the disperse phase concentration and not on the strength of the interactions between particles and continuous phase (Liang et al., 2011). The shear-thinning behaviour found, and its tendency to reach a high-shear-rate limiting viscosity close to the oil medium viscosity, reveal that dispersion microstructure is highly susceptible to shear, yielding a complete disruption at high shear rates (Ten Brinke et al., 2007).

These results clearly point out that both concentration and OC nature determine the rheological properties of the OBM studied, being the influence of both variables on dispersion flow behaviour a key criterion for the mud industry, in order to define downhole conditions, such as penetration rate, adequate viscosity to lift cuttings, hole cleaning, and prevention of excessively high strengths.

## Viscous Flow Behaviour of OBM: Effect of Pressure

The viscous flow curves for B34 and B128 organobentonites dispersions as a function of pressure (1–390 bars), organobentonite nature and disperse phase concentration are shown in Fig. 3. In this sense, Fig. 3A and B gather the viscous flow curves for the lowest organobentonite concentration, whilst Fig. 3C and D collect the ones obtained for the highest organobentonite concentration.

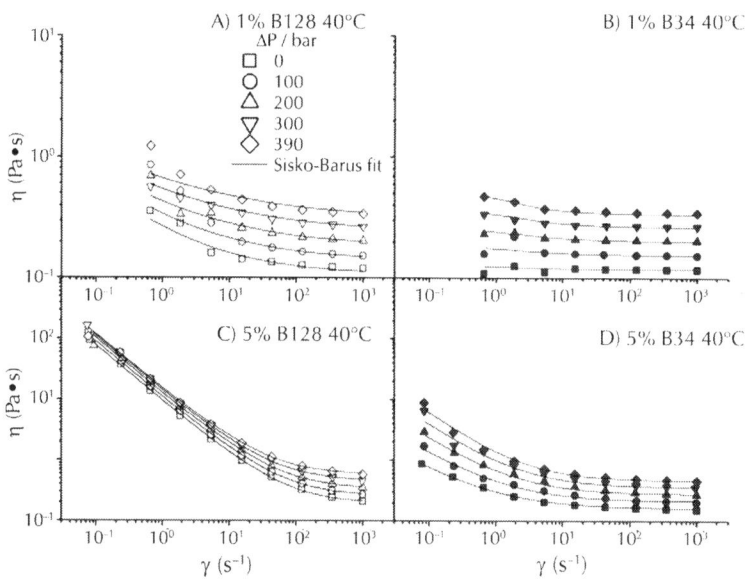

**Figure 3:** Experimental viscous flow curves and Sisko–Barus' model fitting, for the different organobentonite dispersions studied, as a function of pressure, at 40 °C.

As expected, viscosity increases with pressure in the whole range of shear rate tested (Herzhaft et al., 2001). Likewise, the viscous flow behaviour of B34 dispersions is largely modified by an increase in OC concentration, from a quasi-Newtonian behaviour at the lowest concentration (1 m%, Fig. 3B), to an apparent shear-thinning response for the highest concentration (5 m%, Fig. 3D). On the contrary, dispersions made from B128 organobentonite show a clear shear-thinning behaviour in the whole range of disperse phase concentration studied.

In summary, the viscous flow behaviour of these OC dispersions significantly depends on pressure. This dependence can be modelled by using a modified Sisko model that includes the influence of pressure. In this case, Barus› model (Barus, 1893), which satisfactorily describes the pressure dependence of the oil viscosity at constant temperature in this range of pressure, can be used, in combination with Sisko›s model, to take into account both effects in the same equation. The expression, named Sisko–Barus› model, is given as:

$$\eta = \left[\eta_{\infty 0} + k(P)\dot{\gamma}^{n(P)-1}\right] \cdot \exp(\beta(P-P_0)) \qquad (3)$$

Being both consistency, $k\ (P)$, and flow indexes, $n(P)$, linear functions of pressure:

$$k(P) = k_0 + k_1(P-P_0) \qquad (4)$$

$$n(P) = n_0 + n_1(P-P_0) \qquad (5)$$

where $\eta_{\infty 0}$ is the high-shear-rate-limiting viscosity, at the reference pressure and 40 °C; $\beta$ is the piezoviscous coefficient at 40 °C; $P$ is the applied pressure; $k_0$ and $n_0$ are the consistency and flow indexes at the reference pressure ($P_0$ = 1 bar), respectively; and $k_1$, and $n_1$ are fitting parameters (bar$^{-1}$). Sisko–Barus' model fitting parameters have been gathered in Table 2.

As can be seen in Fig. 3A–D, the Sisko–Barus model describes OC dispersion viscosity evolution with both shear rate and pressure fairly well. It is interesting to note that the piezoviscous coefficients of the OBM analysed are quite similar to that of the oil base.

The effect of pressure on the rheology of dispersed system may be attributed to the physical changes that pressure exerts on both continuous and dispersed phases (Combs and Whitmire, 1968 and Hiller, 1963). In this case, OC addition does not seem to significantly affect the bulk piezoviscous properties of the continuous phase, despite the fact that oil phase compression implies a denser particle dispersion, being the effect of pressure on particles negligible as compared to the effect on the continuous oil medium, as has been explained elsewhere (Briscoe et al., 1994). Consequently, continuous phase volume reduction would be the main contribution to the piezoviscous coefficient in these dispersions.

Fig. 4 displays the evolution of the consistency index, for both organophilic bentonite dispersions, with pressure, in the range comprised between 0 and 390 bars, at 40 °C. It is worth remarking that parameter $k_1$ indicates a more or less important change in the consistency index with pressure.

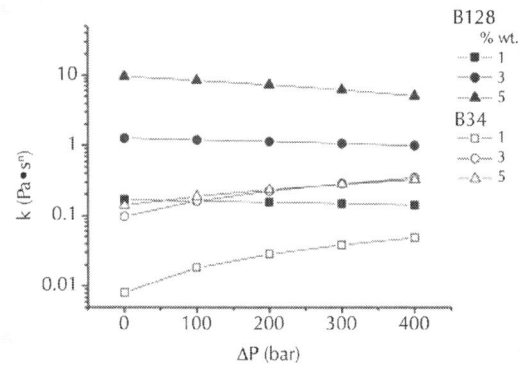

**Figure 4:** Evolution of the consistency index with pressure, as a function of OC nature and concentration (filled symbols: B128 OC dispersions; open symbols: B34 OC dispersions), at 40 °C.

OBM formulated with B34 OC exhibit a similar evolution of the pressure–viscosity curves for OC concentrations above 3 m%, with no significant differences in the consistency indexes, in the range of pressures studied. The positive values of $k_1$ would indicate increasing interparticle interactions, attributed to volume changes in the liquid phase under pressure, resulting in an effective increase in solid concentration.

On the other hand, OBM formulated with B128 always show a decrease of the consistency index with pressure ($k_1 < 0$, see Table 2). This fact suggests that B128 dispersions microstructure would consist of cohesive aggregates, whose strength considerably dampens the influence of oil compression on the bulk rheological response. Probably, a higher compatibility between B128 OC and oil promotes percolated structures, more robust against pressure changes than those developed for B34-based dispersions (Burgentzlé et al., 2004), where a more easily compressible non-packed structure, interconnected by weak interactions, would be developed.

The evolution of the flow index with pressure for the drilling fluid samples studied is shown in Fig. 5. As can be observed, the flow indexes of B34 OC dispersions, at 40 °C, linearly decrease as pressure increases (negative values of the flow index parameter $n_1$ with pressure). Besides, it can be also observed that the most concentrated dispersion (5 m%) shows the highest shear-thinning degree. Changes in the shear-thinning characteristics of dispersions with pressure have been also reported by Alderman et al. (1988), who found that the flow index slightly decreased with pressure, at 40 °C, for WBM, and by Houwen and Geehan (1986), which sustained that, for OBM, the viscous behaviour is largely determined by the physical properties of the continuous phase.

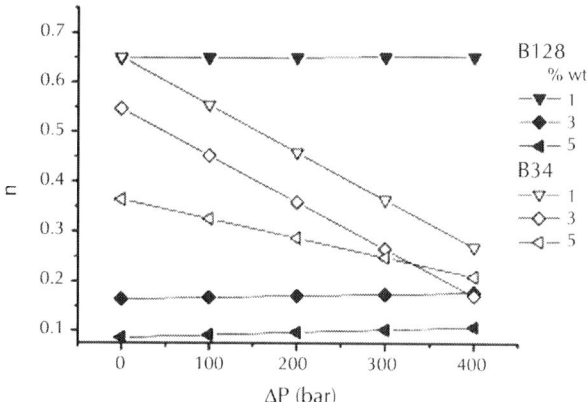

**Figure 5:** Evolution of the flow index with pressure, as a function of OC nature and concentration (filled symbols: B128 OC dispersions; open symbols: B34 OC dispersions), at 40 °C.

For B34-based OBM, the remarkable pressure dependence of the flow index suggests that a higher number of interparticle contacts could be induced by pressure packing. Thus, the dispersion with the largest effective volume fraction has the lowest flow indexes.

On the contrary, B128 dispersions behave completely different, as shown in Fig. 5. Thus, the flow indexes of these dispersions dramatically depend on OC concentration, whilst it is almost independent of pressure. In this case, the flow index is not affected by pressure, since the microstructure would be strong enough to resist changes in pressure. In this sense, the slight influence of pressure on shear-thinning characteristics for B128 dispersions is in agreement with previous results reported by Alderman et al. (1988) for water-based bentonite dispersions, showing that, for strong particle–particle interactions, changes in continuous phase rheology have a small relative influence on the bulk rheology of the dispersion (Briscoe et al., 1994).

The influence of pressure on viscosity, at both intermediate (10 $s^{-1}$) and large (1000 $s^{-1}$) shear rates, and at 40 °C, for selected OBM samples (1% and 5 m %), is illustrated in Fig. 6A–D. As can be observed in Fig. 6C and D, the effect of concentration on fluid piezoviscous coefficient, at high shear rate, does not depend on the type of OC used in the formulation of the fluid, showing very similar piezoviscous coefficients to that of the base oil. On the contrary, in the low shear rate region (Fig. 6A and B), B128 OBM show larger increases in viscosity with OC concentration in relation to B34 OC containing dispersions, although, however, its piezoviscous coefficients decrease as OC concentration increases.

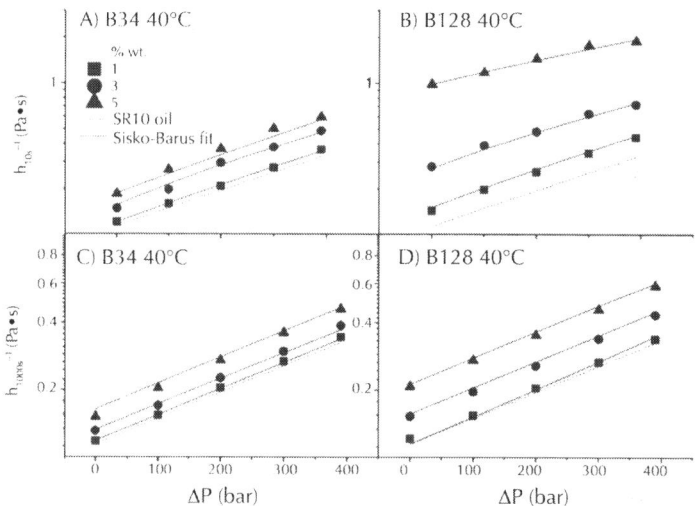

**Figure 6:** Influence of pressure on OBM apparent viscosities, at 40 °C, as a function of OC nature and concentration (symbols correspond to experimental values; solid lines correspond to Sisko–Barus› fitting). A and B: test carried out at 10 s$^{-1}$; C and D: tests carried out at 1000 s$^{-1}$.

On the other hand, the Sisko–Barus model proposed here describes fairly well the pressure dependence of viscosity, for both OC dispersions, at different shear rates. This model can be a useful tool, from the engineering point of view, to calculate pressure losses in the different sections of the bore, as well as being of significant help to solve different problems, i.e. hole cleaning, induced fracturing, and hole erosion during the drilling operation (Abdo and Haneef, 2012, Morita et al., 1996 and Whitfill et al., 2006).

# OBM: Relationship between Dispersion Microstructure and Viscous Flow Behaviour

X-ray diffraction is a powerful technique in order to characterise the structure of clays, OC and composites from a nanoscale point of view (Pavlidou and Papaspyrides, 2008, Ras et al., 2007, Slade and Gates, 2004 and Starodoubtsev et al., 2006). XRD profiles for organobentonite dry powders (B34 and B128) and selected OBM dispersions (5 m% OC) are shown in Fig. 7.

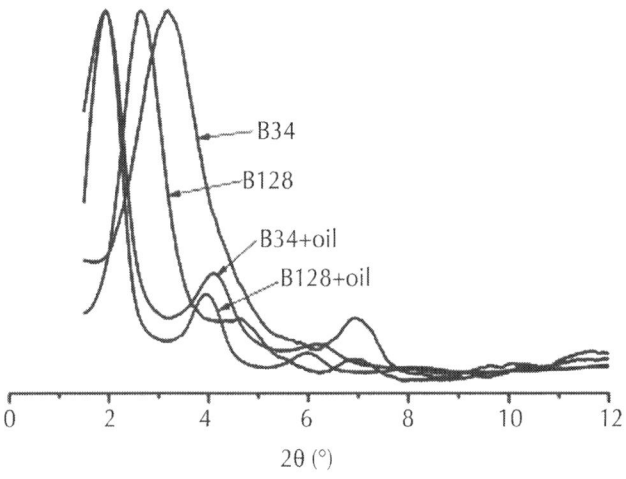

**Figure 7:** XRD patterns of OC dry powders (B34 and B128) and selected OBM (5 m% OC).

The differences observed in the diffraction patterns ($d_{001}$) of both OC (0.577 nm) can be attributed to the different arrangements of organic molecules between the layers. Furthermore, a comparison of these diffractograms with those obtained by Minase et al. (2008) reveals that interlayer structure of alkyl chains strongly depends on the amount of cations adsorbed. Thus, it has been estimated that more than 1.41 mmol of ammonium per gram of clay are present in these organobentonites. In this sense, the interlayer conformations developed by these organic groups could be interpreted as a paraffin-like bilayer structure allowing better fitting of the ammonium groups ( Minase et al., 2008).

The presence of peaks at low angles ($d_{001}$ = 4.506 nm) in both OC dispersions confirms that the oil yields an increase in interlayer spacing. This demonstrates the compatibility of the modified clays with the oil medium. The remarkable increase in interlayer spacing ($d_{001}$) observed for both OC explains the increase in dispersion viscosity (see Fig. 1 and Fig. 2) as compared to base oil viscosity. Similar results have been reported by other authors (Hyun et al., 2001).

Fig. 7 also displays additional peaks. Thus, a second order peak ($d_{002}$) at 1.274 nm, more pronounced for B34 bentonite, and a third order peak, located at 1.430 and 1.479 nm, for B34 and B128 dispersions, respectively. Presumably, the third peak could be related to the formation of monolayer structures. Monolayer structures are possible if the area required for the flat-lying cations, with n carbons atoms in the chain, becomes the same as the area available to each univalent cation in a monolayer between the two silicate layers (Bonczek et al., 2002 and de Paiva et al., 2008). In spite of the fact that both B34 (n-metyl-alkyl ammonium) and B128 (benzyl ammonium) OC show the same maximum swelling peak ($d_{001}$), the second peak ($d_{002}$) presents a slight shifting. This result, in addition of the existence of a third peak corresponding to $d_{003}$, could indicate that the oil creates different disordered structures in both bentonite dispersions, probably due to the partial solvation of hydrocarbons. These materials would form paraffin-like bilayer structures and disordered conformations, such as pseudotrimolecular layers, as has been pointed out elsewhere (He et al., 2004). Additionally, the lower values of second and third order diffraction picks of B128 dispersion could be related to a higher affinity between the aromatic radical group of this OC and the naphthenic nature of the oil ( Burgentzlé et al., 2004 and Paul and Robeson, 2008) and, consequently, yielding a more solid-like configuration of the interlayer structures. Although XRD provides a qualitative structural information for elucidate interlayer spacing, the similar intercalation of the oil molecules observed in both types of dispersions does not explain the different rheological properties observed in these dispersions ( Burgentzlé et al., 2004), suggesting that *d*-spacing swelling parameter is not the more significant variable to predict the viscous flow behaviour.

The swelling of organophilic clays can favour the formation of gels, having a remarkable elasticity, due to several factors, such as the interactions developed between the organophilic ions and the solvent, alkyl chain length and nature of organic groups, and chemical nature of the continuous medium (Gherardi et al., 1996, Jordan, 1949 and Moraru, 2001), among others.

In order to gain some insight into the microstructure–rheology relationship of these OBM, optical micrographs for both B34 (A, B, C) and B128 dispersions (D, F, E) are depicted in Fig. 8. It can be observed that the OBM formulated with B34 OC display a large number aggregated particles dispersed in the continuous oil background. Particle volume fraction and size of these aggregates increase with OC concentration (Wagener and Reisinger, 2003). In contrast, no agglomerated particles can be observed at any concentration for B128 dispersions.

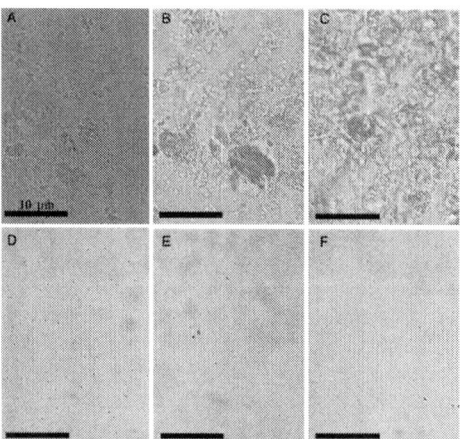

**Figure 8:** Optical micrographs, at 40 °C, for the different OC dispersions studied. A, B, C: B128 OC dispersions (1, 3 and 5 m %); D, E, F: B34 OC dispersions (1, 3 and 5 m %).

The results from XRD and optical microscopy suggest that the microstructure of these dispersions presumably consists of particles and aggregates (King et al., 2007). At the microscopic scale, particles lead to different microstructures depending on the organic ions/solvent interactions. Hence, the dimetyl alkyl covered clay, easily swell, leading to aggregates consisting of multipacked anisotropic shaped particles, probably due to short-range attractive force which prevents disaggregation (Moraru, 2001), not having enough affinity to connect all the aggregates into a percolated assembled structure. Consequently, an increase in OC concentration favours the formation of larger aggregates, as can be observed in Fig. 8, without showing dramatically higher viscosities.

On the contrary, the benzyl ammonium ion in B128 OC yields a significant gel-like behaviour, as a consequence of strong interactions among clay surface and hydrocarbons molecules, due to a better solubility of the benzyl ammonium ion in the naphthenic oil. This OC probably develops a microstructure of tiny aggregates through thin liquid interlayers with consistent linkage points between the tactoids of the physical network (Moraru, 2001), a fact that would explain the higher viscosity values as well as the stronger shear-thinning behaviour of B128 dispersions compared to B34 ones (Krishnamoorti et al., 1996), especially for high OC concentration.

Of course, the above-mentioned microstructure influences dispersion rheology as a function of pressure. For B34 dispersions, the effect of pressure on the rheological parameters ($k$, $n$) of the Sisko–Barus model depends on OC concentration, resulting from the increase in the effective disperse phase volume fraction. In the case of B128 dispersions, the pressure dependence of the viscous behaviour is essentially independent of concentration, due to the significant gel connectivity. However, despite of the different microstructures found for both types of OBM, the effect of compression on viscosity is essentially dominated by the continuous oily phase, showing quite similar piezoviscous values for all the formulations tested.

## CONCLUSIONS

From the experimental results obtained, it can be concluded that the viscous flow behaviour of the OBM investigated is strongly influenced by OC nature and concentration. The pressure–viscosity behaviour of these dispersions is mainly influenced by the piezo-viscous properties of the oil and the properties of the continuous phase. The Sisko–Barus model proposed in the present work predicts the evolution of the non-Newtonian viscosity of oil drillings fluids tested with pressure and shear rate fairly well. This model can be a useful tool, from the engineering point of view, to calculate pressure losses in the different sections of the bore, as well as being of significant help to solve other additional problems, such as hole cleaning, induced fracturing, and hole erosion during the drilling operation.

In spite of the fact that both OC display many similarities at the nanoscale level, the microstructures resulting from interactions among organic ions and solvent are quite diverse, yielding significant differences in the bulk viscous flow behaviour of their respective dispersions.

## ACKNOWLEDGEMENTS

This work has been sponsored by FEDER-Excellence Projects Programme (Research project P08-TEP-3895, Junta de Andalucía, Spain). The authors gratefully acknowledge the financial support.

## REFERENCES

1. Abdo, J., Haneef, M.D., 2012. Nano-enhanced drilling fluids: pioneering approach to overcome uncompromising drilling problems. J. Energy Resour. Technol. 134, 014501. http://dx.doi.org/10.1115/1.4005244.

2. Alderman, N.J., Gavignet, A., Guillot, D., Maitland, G.C., 1988. High-temperature, highpressure rheology of water based muds. Society of Petroleum Engineers of AIME, (Paper), 18035, pp. 187–195.
3. Barus, C., 1893. Isothermals, isopiestics and isometrics relative to viscosity. Am. J. Sci. 45, 87–96.
4. Bonczek, J.L., Harris, W.G., Nkedi-Kizza, P., 2002. Monolayer to bilayer transitional arrangements of hexadecyltrimethylammonium cations on Na-Montmorillonite. Clay Clay Miner. 50, 11–17.
5. Briscoe, B.J., Luckham, P.F., Ren, S.R., 1994. The properties of drilling fluids at high pressure and high temperatures. Philos. Trans. R. Soc. Lond. A 348, 179–207.
6. Burgentzlé, D., Duchet, J., Gérard, J.F., Jupin, A., Fillon, B., 2004. Solvent-based nanocomposite coatings I. Dispesions of oganophilic montmorillonita in organic solvents. J. Colloid Interface Sci. 278, 26–39.
7. Caenn, R., Chillingar, G.V., 1996. Drillings fluids: state of art. J. Pet. Sci. Eng. 14, 221–230.
8. Chilingarian, G.V., Vorabutr, P., 1983. Drilling and Drilling Fluids, Second ed. Elsevier, Amsterdam, Netherlands.
9. Combs, G.D., Whitmire, L.D., 1968. Capillary viscometer simulates bottom-hole conditions. Oil Gas J. 108–113 (30 September).
10. Darley, H.C.H., Gray, G.R., 1988. Composition and properties of drilling and completion fluids, Sixth ed. Gulf Publ. Co., Houston, USA.
11. Das Kanungo, J.L., McAtee Jr., J.L., 1986. Effects of polymers and clay concentrations on the viscosities of organo-smectite dispersions under high pressure. Appl. Clay Sci. 1, 285–293.
12. de Paiva, L.B., Morales, A.R., Velenzuela Díaz, F.R., 2008. Organoclays: properties, preparation and applications. Appl. Clay Sci. 42, 8–24.

13. England, K.W., Parris, M.D., 2010. Viscosity influences of high pressure on borate crosslinked gels. SPE Deepwater Drilling and Completions Conference, Galveston, USA.
14. Gandelman, R.A., Leal, R.A.F., Gonçalves, Aragão, A.F., Lomba, R.F., Martins, A.L., 2007. Study on gelation and freezing phenomena of synthetic drilling fluids in ultradeepwater environments. SPE/IADC Drilling Conference and Exhibition, Amsterdam, Netherlands.
15. Ghalambor, A., Ashrafizadeh, S.N., Nasiri, M., 2008. Effect of basic parameters on the viscosity of synthetic-based drilling fluids. SPE International Symposium and Exhibition on Formation Damage Control, Lafayette, USA.
16. Gherardi, B., Tahani, A., Levitz, P., Bergaya, F., 1996. Sol/gel phase diagrams of industrial organo-bentones in organic media. Appl. Clay Sci. 11, 163–170.
17. Hato, M.J., Zhang, K., Ray, S.S., Choi, H.J., 2011. Rheology of organoclay suspension. Colloid Polym. Sci. 289, 1119–1125.
18. Hauser, E.A., 1950. Modified gel-forming clay and process of producing same. U.S. Patent No. 2,531,427.
19. He, H., Frost, R.L., Deng, F., Zhu, J., Wen, X., Yuan, P., 2004. Conformation of surfactant molecules in the interlayer of montmorillonite studied by 13C MAS NMR. Clay Clay Miner. 52, 350–356.
20. Herzhaft, B., Peysson, Y., Isambourg, P., Delepoulle, A., Toure, A., 2001. Rheological properties of drilling muds in deep offshore conditions. SPE/IADC Drilling Conference, Amsterdam, Netherlands.
21. Hiller, K.H., 1963. Rheological measurements of clay suspensions at high temperatures and pressures. J. Petrol. Technol. 15, 779–788.
22. Houwen, O.H., Geehan, T., 1986. Rheology of Oil-Base Muds. SPE Annual Technical Conference and Exhibition, New Orleans, USA.

23. Hyun, Y.H., Lim, S.T., Choi, H.J., Jhon, M.S., 2001. Rheology of poly(ethylene oxide)/ organoclay nanocomposites. Macromolecules 34, 8084–8093.
24. Jordan, J.W., 1949. Organophilic bentonites. I. Swelling in organic liquids. J. Phys. Chem. 53, 294–306.
25. Jordan, J.W., Nevins, M.J., Stearns, R.C., Cowan, J.C., Beasley, A.E., 1965. Well- working fluids. U.S. Patent No. 3,168,475.
26. Khodja, M., Canselier, J.P., Bergaya, F., Fourar, K., Khodja, M., Cohaut, N., Benmounah, A., 2010. Shale problems and water-based drilling fluid optimisation in the Hassi Messaoud Algerian oil field. Appl. Clay Sci. 49, 383–393.
27. King Jr., H.E., Milner, S.T., Lin, M.Y., Singh, J.P., Mason, T.G., 2007. Structure and rheology of organoclay suspensions. Phys. Rev. E. 75, 021403-1–021403-20.
28. Krishnamoorti, R., Vaia, R.A., Giannelis, E.P., 1996. Structure and dynamics of polymerlayered silicate nanocomposites. Chem. Mater. 8, 1728–1734.
29. Le Pluart, L., Duchet, J., Sautereau, H., Halley, P., Gerard, J.F., 2004. Rheological properties of organoclay suspensions in epoxy network precursors. Appl. Clay Sci. 25, 207–219.
30. Liang, R., Han, L., Doraiswamy, D., Gupta, R.K., 2011. The rheology of aramid platelet suspensions. Polym. Eng. Sci. 51, 1933–1941.
31. Massinga, P.H., Focke, W.W., de Vaal, P.L., Atanasova, M., 2010. Alkyl ammonium intercalation of Mozambican bentonite. Appl. Clay Sci. 49, 142–148.
32. Menezes, R.R., Marques, L.N., Campos, L.A., Ferreira, H.S., Santana, L.N.L., Neves, G.A., 2010. Use of statistical design to study the influence of CMC on the rheological properties of bentonite dispersions for water-based drilling fluids. Appl. Clay Sci. 49, 13–20.
33. Meng, X., Zhang, Y., Zhou, F., Chu, P.K., 2012. Effects of carbon ash on rheological properties of water drilling fluids. J. Pet. Sci. Eng. 100, 1–8.

34. Minase, M., Kondo, M., Onikata, M., Kawamura, K., 2008. The viscosity of organic liquid suspensions of trimethyldococylammonium-montmorillonite complexes. Clay Clay Miner. 56, 49–65.
35. Moraru, V.N., 2001. Structure formation of alkylammonium montmorillonites in organic media. Appl. Clay Sci. 19, 11–26.
36. Morita, N., Black, A.D., Fuh, G.-F., 1996. Borehole breakdown pressure with drilling fluidsI. Empirical results. Int. J. Rock Mech. Min. Sci. Geomech. Abstr. 33, 39–51.
37. Mueller, S., Llewellin, E.W., Mader, H.M., 2010. The rheology of suspensions of solid particles. Proc. R. Soc. A 466, 1201–1228.
38. Paul, D.R., Robeson, L.M., 2008. Polymer nanotechnology: Nanocomposites. Polymer 49, 3187–3204.
39. Pavlidou, S., Papaspyrides, C.D., 2008. A review on polymer-layered silicate nanocomposites. Prog. Polym. Sci. 33, 1119–1198.
40. Politte, M.D., 1985. Invert oil mud rheology as a function of temperature and pressure. SPE/IADC Drilling Conference, New Orleans, USA.
41. Ras, R.H.A., Umemura, Y., Johnston, C.T., Yamagishi, A., Schoonheydt, R.A., 2007. Ultrathin hybrid films of clay minerals. Phys. Chem. Chem. Phys. 9, 918–932.
42. Santoyo, E., Santoyo-Gutiérrez, S., García, A., Espinosa, G., Moya, S.L., 2001. Rheological property measurement of drilling fluids used in geothermal wells. Appl. Therm. Eng. 21, 283–302.
43. Slade, P.G., Gates, W.P., 2004. The swelling of HDTMA smectitec as influenced by their preparation and layer charges. Appl. Clay Sci. 25, 93–101.
44. Starodoubtsev, S.G., Lavrentyeva, E.K., Khokhlov, A.R., Allegra, G., Famulari, A., Meille, S.V., 2006. Mechanism of smectic arrangement of montmorillonite and bentonite clay platelets incorporated in gels of poly(acrylamide) induced by

the interaction with cationic surfactants. Langmuir 22, 369–374.

45. Ten Brinke, A.J.W., Bailey, L., Lekkerkerker, H.N.W., Maitland, G.C., 2007. Rheology modification in mixed shape colloidal dispersions. Part I: Pure components. Soft Matter 3, 1145–1162.

46. Tropea, C., Yarin, A.L., Foss, J.F., 2007. Springer handbook of experimental fluid mechanics, First ed. Springer Verlag, Berlin, Germany.

47. Turian, R.M., Ma, T.-W., Hsu, F.-L.G., Sung, D.-J., 1998. Flow of concentrated nonNewtonian slurries: 1. Friction losses in laminar, turbulent and transition flow through straight pipe. Int. J. Multiphase Flow 24, 225–242.

48. Turian, R.M., Attal, J.F., Sung, D.-J., Wedgewood, L.E., 2002. Properties and rheology of coalwater mixtures using different coals. Fuel 81, 2019–2033.

49. Wagener, R., Reisinger, T.J.G., 2003. A rheological method to compare the degree of exfoliation of nanocomposites. Polymer 44, 7513–7518.

50. Wang, F., Wang, R., Liu, J., Wang, L., Li, J., Che, L., Su, H., 2010. Rheology of high-density water-based drilling fluid at high temperature and high pressure. Shiyou Xuebao/ Acta Petrol. Sin. 31, 306–310.

51. Wang, J., Zheng, J., Musa, O.M., Farrar, D., Cockcroft, B., Robinson, A., Gibbison, R., 2011. Salt-tolerant, thermally-stable rheology modifier for oilfield drilling applications. SPE International Symposium on Oilfield Chemistry, The Woodlands, USA.

52. Weir, I.S., Bailey, W.J., 1996. Statistical study of rheological models for drilling fluids. SPE J. 1, 473–485.

53. Whitfill, D.L., Jamison, D.E., Wang, M., Thaemlitz, C., 2006. New design models and materials provide engineered solutions to lost circulation. SPE Russian Oil and Gas Technical Conference and Exhibition, Moscow, Russia.

54. Zhang, L.-M., Jahns, C., Hsiao, B.S., Chu, B., 2003. Synchotron SAXS/WAXD and rheological studies of clay suspensions in silicone fluid. J. Colloid Interface Sci. 266, 339–345.

# Citations

## CHAPTER 1

Yuanhua Lin, Xiangwei Kong, Yijie Qiu, and Qiji Yuan, "Calculation Analysis of Pressure Wave Velocity in Gas and Drilling Mud Two-Phase Fluid in Annulus during Drilling Operations," Mathematical Problems in Engineering, vol. 2013, Article ID 318912, 17 pages, 2013. doi:10.1155/2013/318912.

## CHAPTER 2

Gilbert Makanda, Sachin Shaw, and Precious Sibanda, "Diffusion of Chemically Reactive Species in Casson Fluid Flow over an

Unsteady Stretching Surface in Porous Medium in the Presence of a Magnetic Field,"Mathematical Problems in Engineering, Article ID 724596, in press.

# CHAPTER 3

Linjing Zhu, Hongqiao Lan, Bingjing He, Wei Hong, and Jun Li, "Encapsulation of Menthol in Beeswax by a Supercritical Fluid Technique," International Journal of Chemical Engineering, vol. 2010, Article ID 608680, 7 pages, 2010. doi:10.1155/2010/608680.

# CHAPTER 4

Mondala, A. , Al-Mubarak, R. , Atkinson, J. , Shields, S. , Young, B. , Senger, Y. and Pekarovic, J. (2015) Direct Solid-State Fermentation of Soybean Processing Residues for the Production of Fungal Chitosan by Mucor rouxii.Journal of Materials Science and Chemical Engineering, 3, 11-21. doi: 10.4236/msce.2015.32003.

# CHAPTER 5

Zhi-Lei Zhang, Feng-Shan Zhou, Yi-He Zhang, et al., "A Promising Material by Using Residue Waste from Bisphenol A Manufacturing to Prepare Fluid-Loss-Control Additive in Oil Well Drilling Fluid," Journal of Spectroscopy, vol. 2013, Article ID 370325, 10 pages, 2013. doi:10.1155/2013/370325.

# CHAPTER 6

Mehbad, N. (2014) Evaluation of Additives as Corrosion Inhibitors/Antioxidants for High Quality Nano Emulsifiable Oils of Metalworking Fluids. Journal of Surface Engineered Materials and Advanced Technology, 4, 155-163. doi:10.4236/jsemat.2014.43019.

# CHAPTER 7

G. Sozhamannan, S. Prabu and V. Venkatagalapathy, "Effect of Processing Paramters on Metal Matrix Composites: Stir Casting Process," Journal of Surface Engineered Materials and Advanced Technology, Vol. 2 No. 1, 2012, pp. 11-15. doi: 10.4236/jsemat.2012.21002.

# CHAPTER 8

Ivica Kisic, Sanja Mesic, Ferdo Basic, Vladislav Brkic, Milan Mesic, Goran Durn, Zeljka Zgorelec, Lidija Bertovic, The effect of drilling fluids and crude oil on some chemical characteristics of soil and crops, Geoderma, Volume 149, Issues 3–4, 15 March 2009, Pages 209-216, ISSN 0016-7061, http://dx.doi.org/10.1016/j.geoderma.2008.11.041.

# CHAPTER 9

J. Hermoso, F. Martinez-Boza, C. Gallegos, Influence of viscosity modifier nature and concentration on the viscous flow behaviour of oil-based drilling fluids at high pressure, Applied Clay Science, Volume 87, January 2014, Pages 14-21, ISSN 0169-1317, http://dx.doi.org/10.1016/j.clay.2013.10.011.

# Index

### A
Alkali insoluble materials (AIMs) 103

### B
Bisphenol A (BPA) 120
Bottomhole pressure (BHP) 3

### C
Critical micelle concentration (CMC) 149

### D
Degree of deacetylation (DDA) 98, 104, 113

### E
Equation of state (EOS) 12
Equations of state (EOS) 8
European Waste Catalogue (EWC) 121

### F
Foundry casting processes 165
Fourier transfer infrared (FTIR) 104
Fourier transform infrared (FTIR) 98, 120

### H
Heavy metals (HM) 178

## M

Magnetite nanoparticles (MNPs) 79
Managed pressure drilling (MPD) 2
Matrix material 166
Menthol mass fraction 78, 82, 83, 92, 93, 94
Metalworking 161, 236
Mineral oils (MO) 178, 187

## N

Nuclear magnetic resonance (NMR) 120

## O

Oil based muds (OBM) 208
Organic matter (OM) 178
Orthogonal array (OA) 127

## P

Particles from gas-saturated solutions (PGSS) 77
Polycyclic aromatic hydrocarbons (PAHs) 178, 188
Potato dextrose agar (PDA) 101

## R

Rotating Control Device (RCD) 26
Runge-Kutta method (R-K4) 22

## S

Separation-condensation method 78
Solid-state fermentation (SSF) 97, 100, 101
Successive linearization method (SLM) 57

## T

Total petroleum hydrocarbons (TPH) 178, 191

## V

Volume fraction 169

## W

Water-based muds (WBM) 209
World Health Organization's (WHO) 192